"With a keen eye and grim wit, McQuaig's perceptive inquiry into the world's energy system strips away layer after layer of deceit, cynicism, racism, sordid manipulation, violence and aggression, in the dedicated effort to extract every possible ounce of profit and power in a race to the edge of disaster, perhaps beyond. It is an urgent wake-up call that should—that must—be read and acted upon, without delay."

—Noam Chomsky

"Intriguing. . . . McQuaig keeps her critiques intelligent. . . . [She] hits the nail on the head when she tackles the question of why the United States is so concerned about oil." —*The Globe and Mail*

"McQuaig gives the reader an entertaining crash course on the history of the oil industry. . . . It's a highly educational rant . . . and a deliciously written one." —*The Gazette* (Montreal)

"Extremely interesting. . . . McQuaig writes extremely well, and has clearly done her homework." —*Winnipeg Free Press*

"An analysis presented with passion and impressive detail."—*Maclean's*

"This is an essential book, a call to action, written with passion and wit." —*Vancouver Review*

"McQuaig catches a system at work and offers a perspective on the United States' mechanism—military and commercial—for controlling and squandering the earth's resources. . . . McQuaig has been calling our attention to instances of power run amok for decades now; this time, the notice seems final." —*The Vancouver Sun*

"Linda McQuaig's best book yet . . . a story of corporate greed, short-sightedness and business as usual." —*Kitchener-Waterloo Record*

"McQuaig's book certainly demonstrates courage in the way she writes with such conviction, especially considering the general reluctance of the popular media to make the broad claims she does. . . . [A] satisfying and timely read." —*The Hamilton Spectator*

"Linda McQuaig's best book yet. . . . McQuaig's is a crusading voice." —*The Guelph Mercury*

WAR, BIG
OIL, AND

ALSO BY LINDA McQUAIG

All You Can Eat: Greed, Lust and the New Capitalism
*The Cult of Impotence: Selling the Myth of Powerlessness in the
    Global Economy*
Shooting the Hippo: Death by Deficit and Other Canadian Myths
The Wealthy Banker's Wife: The Assault on Equality in Canada
*The Quick and the Dead: Brian Mulroney, Big Business and
    the Seduction of Canada*
*Behind Closed Doors: How the Rich Won Control of Canada's Tax
    System—and Ended Up Richer*

ANCHOR CANADA

LINDA McQUAIG

THE FIGHT FOR THE PLANET

IT'S THE CRUDE, DUDE

LIBRARY AND ARCHIVES CANADA CATALOGUING IN PUBLICATION

McQuaig, Linda, 1951–
War, big oil, and the fight for the planet / Linda McQuaig.—Updated
Anchor Canada ed.

Previously published under title: It's the crude, dude.
Includes bibliographical references and index.

ISBN-13: 978-0-385-66347-2
ISBN-10: 0-385-66347-1

1. Petroleum industry and trade—Political aspects. 2. Iraq War,
2003-. 3. Petroleum industry and trade—Environmental aspects.
4. Global warming. 5. Fossil fuels—Environmental aspects. I. Title.

HD9560.5.M38 2006   333.8'232   C2006-903160-6

Printed and bound in Canada

Published in Canada by
Anchor Canada, a division of
Random House of Canada Limited

Visit Random House of Canada Limited's website:
www.randomhouse.ca

TRANS 10 9 8 7 6 5 4 3 2 1

To my father, Jack McQuaig,
whose love of writing got me writing.

And my mother, Audrey McQuaig,
who, despite five children, made me feel special.

And, as ever, to my precious Amy.

CONTENTS

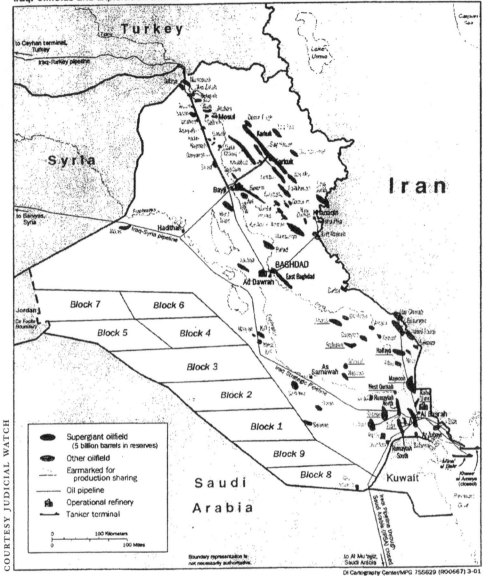

Iraqi Oilfields and Exploration Blocks

35AS0713

Map of Iraq studied by Cheney task force on energy,
released under U.S. court order.

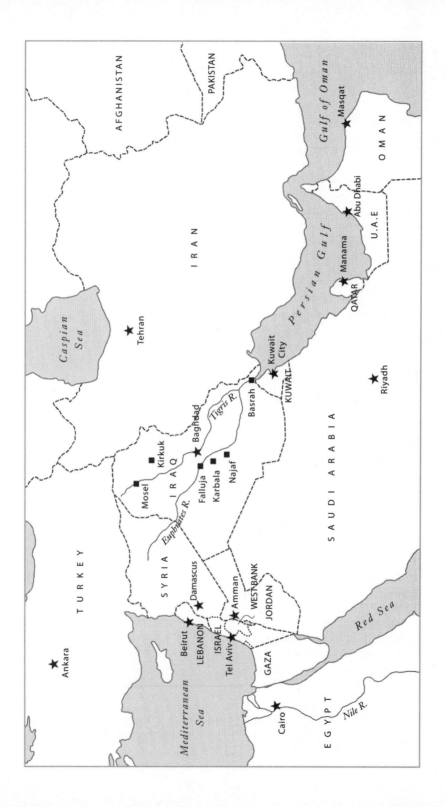

TIMELINE

Circa 500,000,000 B.C.—Organic materials, transformed under heat and pressure, are buried in the earth's crust, forming vast reservoirs of fossil fuels: oil, coal and gas.

1820s—French scientist Jean-Baptiste-Joseph Fourier figures out that some of the sun's rays, after bouncing back from the earth, are trapped in the atmosphere, thereby warming the earth.

1850s—The possibility of crude oil from the ground being transformed into a potent energy source is explored.

1857—Oil is discovered in Oil Springs, Ontario.

1859—Oil is discovered in Titusville, Pennsylvania.

1862—John D. Rockefeller invests in his first oil refinery in Cleveland, Ohio, and soon becomes the dominant player in the U.S. oil industry.

1895—Swedish chemist Svante August Arrhenius predicts that the excessive burning of fossil fuels by humans will eventually alter the earth's climate.

1901—The first deal to develop oil in the Middle East is signed between the Shah of Persia (later Iran) and a London-based financier who establishes the Anglo-Persian Oil Company (which later becomes BP).

1911—The U.S. Supreme Court orders the breakup of Rockefeller's Standard Oil empire. Rockefeller interests continue to dominate the oil industry through a company now known as Exxon.

1925—The newly created government of Iraq signs a deal granting a consortium of European and American oil interests (later known as the Iraq Petroleum Company) a concession to develop Iraq's oil until the year 2000.

1928—The heads of the three dominant international oil companies—Exxon, BP and Shell—meet at Achnacarry Castle in Scotland to carve up the international oil market amongst themselves.

1938—Oil is discovered in Saudi Arabia, and is eventually developed by Aramco, a consortium of U.S. oil companies, including Exxon.

1945—U.S. president Franklin D. Roosevelt meets Saudi Arabia's King Abdul-Aziz ibn Saud and establishes a special understanding between the U.S. and Saudi Arabia.

1948—Venezuela improves its deal with oil companies, under which it receives a 50 percent royalty. (Saudi Arabia is only receiving a 12 percent royalty on its oil.)

1951—The democratically elected government of Mohammed Mossadegh nationalizes Iran's oil industry. In protest, the major international oil companies organize a worldwide boycott of Iran's oil. The Iranian economy is crippled by the boycott, but the government refuses to back down.

1953—Washington sends CIA operative Kermit Roosevelt to Tehran, where he orchestrates a coup that overthrows the Mossadegh government. The Shah is installed as Iran's absolute monarch, with the support of Washington.

1956—Geophysicist M. King Hubbert predicts that oil production in the U.S. will peak in 1970.

1958—Brigadier General Abdul Qarim Qasim takes power in Iraq and unsuccessfully pressures foreign oil interests to give the country a stake in its domestic oil industry.

1959—Washington imposes restrictions on foreign oil imports—restrictions that remain in place for the next fourteen years, contributing to the depletion of U.S. oil reserves and the eventual U.S. dependence on Middle Eastern oil.

1960—The Organization of Petroleum Exporting Countries (OPEC) is established at a meeting in Baghdad, largely as a result of the efforts of Abdullah Tariki of Saudi Arabia and Juan Pablo Pérez Alfonzo of Venezuela.

   Tariki is appointed Saudi Arabia's oil minister in a reformist cabinet that includes prominent advocates of democratic reform and greater Saudi control over the kingdom's oil. The reformist experiment is quickly brought to an end.

1962—Tariki is replaced as Saudi oil minister by pro-Western sheikh Zaki Yamani.

1963—In Iraq, the popular Qasim government is overthrown by the more extremist Baathist Party.

1968—A second coup in Iraq enables the Baathist Party to consolidate control, bringing Saddam Hussein to power.

1969—A young army officer from a nomadic background, Mu'ammer al Qaddafi, seizes power in Libya.

1970—Qaddafi negotiates a significantly better deal for the country with the international oil industry.

   U.S. oil production peaks, as predicted by M. King Hubbert.

1972—Iraq nationalizes its oil industry, beginning a wave of oil nationalizations in the Middle East.

1973—The Arab-Israeli War breaks out.

   For the first time, OPEC unilaterally raises the price of oil. Within a few months, it quadruples the price. The revenues of OPEC nations soar.

   The Arab nations of OPEC cut their oil production to increase pressure for an Israeli withdrawal from Palestinian territories.

   Saudi Arabia cuts off all oil shipments to the U.S.

1974—The Arab-Israeli War ends, and the five-month-long Arab oil embargo is called off.

   Exxon reports a profit for the previous year of $2.5 billion—the biggest profit ever reported by a corporation.

1975—The U.S. introduces new regulations (known as CAFE standards) requiring automakers to make their cars more fuel-efficient. Significant improvement in fuel efficiency is quickly achieved.

At a summit in Algiers, radicals within OPEC, led by Algeria, try to push the organization into becoming "the shield of the Third World." The attempt ultimately fails.

1979—In Iran, a popular uprising, led by Islamic militants, overthrows the shah and takes control. A group of militants storm the U.S. embassy in Tehran and seize fifty-two hostages in a drama that drags on for fourteen months.

1980—In what becomes known as the Carter doctrine, U.S. president Jimmy Carter declares that any attempt by an outside force to gain control of the Persian Gulf region will be regarded as an assault on U.S. vital interests and "repelled by any means necessary, including military force."

1988—An international scientific conference on global warming is held in Toronto, and its report concludes that "humanity is conducting an unintended, uncontrolled, globally pervasive experiment whose ultimate consequences are second only to global nuclear war."

The United Nations and the World Meteorological Organization establish the Intergovernmental Panel on Climate Change (IPCC) to assess the scientific understanding of global warming.

1990—Iraq invades Kuwait.

1991—U.S.-led coalition forces drive Iraq out of Kuwait and decisively defeat the Iraqi army.

1992—At the "Earth Summit" in Rio de Janeiro, leaders of 154 countries, including the U.S. and Canada, sign a legally binding convention committing themselves to address global warming.

1993—Canada, the U.S. and Mexico sign the North American Free Trade Agreement (NAFTA), under which Canada agrees to continental energy-sharing, long sought by Washington. Mexico refuses to sign this part of the agreement.

1995—After input from scientists around the world, an IPCC report makes the case that the earth's climate is in danger of being altered from the buildup of greenhouse gases in the atmosphere, and that human activities are contributing to the buildup.

Dick Cheney becomes CEO of oil services giant Halliburton.

1997—The "Project for the New American Century" (PNAC) is established by a group of U.S. neoconservatives, including Dick Cheney, Donald Rumsfeld, Paul Wolfowitz and others who will later become key figures in the administration of George W. Bush.

World leaders meet in Kyoto, Japan, and establish a

timetable and targets for the reduction of greenhouse gas emissions around the world.

1998—The PNAC writes a letter to President Bill Clinton urging him to make the removal of Saddam Hussein a U.S. objective. The letter refers to "our vital interests in the Gulf" which, it says, include "a significant portion of the world's supply of oil."

2000—Venezuelan president Hugo Chávez "saves" OPEC and argues it should serve as a model for Third World development.

A major PNAC paper urges that the U.S. be transformed into "tomorrow's dominant force," but argues that this could take a long time in the absence of "some catastrophic and catalyzing event—like a new Pearl Harbor."

2001—At the first meeting of its National Security Council, top officials of the new Bush administration discuss military action against Saddam Hussein.

A U.S. task force on energy, chaired by vice-president Dick Cheney, is quickly set up and meets amid great secrecy. It considers the "capture of new and existing oil and gas fields."

In March, the U.S. announces it will not ratify Kyoto.

On September 11, terrorists connected to Al Qaeda strike the World Trade Center and the Pentagon, killing close to 3,000 people. The next day, President Bush is focused on establishing a link between Saddam Hussein and Al Qaeda.

2002—In October, members of Cheney's staff meet with executives from major U.S. oil companies, including Halliburton, according to *The Wall Street Journal.*

Philip Carroll, former CEO of Shell Oil, is retained by the Pentagon to draft a strategy for developing Iraq's oil sector by setting up a U.S.-style corporation to oversee it.

A subsidiary of Halliburton is retained by the Pentagon to draw up plans for putting out oil fires in the event of a war in Iraq.

In December, Canada ratifies Kyoto.

2003—In February U.S. secretary of state Colin Powell appears before the United Nations to urge UN action against Iraq because of the dangers posed by Iraq's "weapons of mass destruction." In the same month, a confidential U.S. government document outlines plans to replace Iraq's largely state-run economy with a mostly privatized economy. The plans include the privatization of Iraq's oil sector.

In March, a $7-billion Pentagon contract for putting out oil fires is awarded to a Halliburton subsidiary. (Halliburton's Iraq-related contracts eventually rise to $18 billion.)

On March 20, the U.S.-led "coalition" invades Iraq and

quickly overwhelms the Iraqi
army. It takes control of
Baghdad by early April.

2004—Top U.S. weapons inspector
David Kay testifies in January
that, despite months of searching
inside Iraq, no evidence can be
found that Iraq had "weapons of
mass destruction."

In June, the National
Commission on Terrorist
Attacks upon the United States
(known as the 9/11 commission)
releases a comprehensive interim
report in which it debunks Bush
administration claims that
Saddam Hussein had links to
Al Qaeda.

2005—In January, millions of Iraqis
vote in a national election. The
one platform common to all
parties running is the need to
end the U.S. occupation.

In February, a report by the
U.S. Department of Energy's
National Energy Laboratory
describes the scope of the
problem the world will face
when global oil production
peaks: "The world has never
faced a problem like this . . .
Previous transitions (wood to
coal and coal to oil) were gradual
and evolutionary; oil peaking will
be abrupt and revolutionary."

In October, Col. Lawrence
Wilkerson, former chief of staff
to Colin Powell, describes how
focused the Bush administration
was on gaining control of Middle
Eastern oil. During a question-
and-answer period at a policy

forum of the New America
Foundation, Wilkerson says:
"We had a discussion in policy
planning about actually mounting
an operation to take the oilfields
in the Middle East, internationalize them, put them under some
sort of UN trusteeship and
administer the revenues and the
oil accordingly. That's how
serious we thought about it."

2006—At the annual meeting of
the American Economic
Association, Nobel prize-
winning economist Joseph
Stiglitz presents a new estimate
of the likely cost of the war in
Iraq: between $1 trillion to $2
trillion.

In February, Republican
Congressman Joe Barton of
Texas launches an anti-trust
investigation into Citgo, a
Venezuelan-owned oil company.
Barton's investigation is to
examine Citgo for its policy of
providing poor communities in
the U.S. with heating oil at
below-market prices.

In March, the *New York
Times* reports on a confidential
British government memo which
appears to show that George W.
Bush was intent on invading Iraq
months before the invasion
began, whether or not international arms inspectors found
weapons of mass destruction.
The memo stated: "The start
date for the military campaign
was pencilled in for 10 March."

PREFACE

The legendary doyenne of the White House press corps minced no words when she finally got her chance in March 2006 to put a question to the President: "Every reason given [for invading Iraq] has turned out not to be true . . . why did you really want to go to war?" asked Helen Thomas. "You have said it wasn't oil . . . or anything else. What was it?"

George W. Bush sidestepped Thomas' question about his true motive for war, insisting instead that he didn't really want to go to war, that "no President wants war."

The little dust-up between the President and the feisty octogenarian attracted a brief flurry of media attention. But there was no accompanying media attention to Thomas' compelling question, which simply went unanswered, as it has for years.

Fresh evidence keeps coming to light—including revelations from top-secret British memos—about the intensity of the Bush administration's determination to invade Iraq, whether or not Iraq had weapons of mass destruction. Yet we don't insist on knowing "why?"

Any suggestion of a possible oil motive in the war is still routinely dismissed as the terrain of conspiracy theorists.

The fact that Iraq is the last easily-harvested oil bonanza left on earth—a vast, virtually untouched reservoir of the world's most valuable resource—is largely ignored, as pundits focus instead on the administration's professed concern about building democracy in Iraq.

The refusal to take seriously the possibility of an oil motive is bizarre, given oil's obvious geopolitical significance and the intense focus on oil exhibited by U.S. administrations, particularly the current one. George W. Bush himself highlighted the problem of America's dependence on foreign oil in his 2006 State of the Union address.

And in a surprisingly frank comment in October 2005, Col. Lawrence Wilkerson, former chief of staff to Colin Powell, told a policy forum of the New America Foundation: "We had a discussion in policy planning about actually mounting an operation to take the oilfields in the Middle East, internationalize them, put them under some sort of UN trusteeship and administer the revenues and the oil accordingly. That's how serious we thought about it."

Even this intriguing comment, from a high-level administration insider, failed to provoke media questioning about a possible oil motive.

The centrality of oil to the modern world is well known. Oil is integrally related to virtually every aspect of our way of life from transportation, communication and the mass production of goods, to food, heat, light and military power. Access to oil is therefore essential to modern living, as well as being crucial to maintaining military dominance—as was amply demonstrated in both World Wars.

If we separate the reality of oil's importance from the politically sensitive issue of Iraq, we can easily appreciate why getting control of oil reserves has for decades been an important goal of powerful nations. This is particularly true of the United States, which is keen to protect its position as the dominant global power with the world's most advanced standard of living.

The problem the U.S. faces is that while its appetite for oil is virtually unlimited, its reserves are *quite* limited. It consumes roughly 25 percent of all the oil produced in the world each year, but has only 3 percent of the world's crude reserves. To make up for the shortfall, the U.S. relies heavily on oil from outside its borders, leaving it vulnerable if key reserves are under the control of hostile nations. Overcoming this vulnerability has been a central goal of U.S. policymakers, particularly since Arab nations dramatically cut back their oil exports for a brief period in the early 1970s.

With roughly two thirds of the world's oil located in the Middle East, a major thrust of U.S. policy has long been gaining control over the region and its reserves.

Thus, for the past sixty years, Washington has provided crucial backing for the dictatorship in Saudi Arabia, protecting the royal family against threats from inside and outside the country. In exchange, Washington (along with U.S. oil companies) has enjoyed extraordinary leverage over Saudi oil policy, which effectively amounts to leverage over world oil policy.

Washington's intervention in Iran has been even more direct. After a popular nationalist leader was elected there in the early 1950s, Tehran nationalized its foreign-owned oil industry. The big British and American oil companies were so incensed that they organized a worldwide boycott of nationalized Iranian oil. When the boycott failed to bring Tehran to its knees, Washington orchestrated a coup in 1953 that toppled the fledgling Iranian democracy and replaced it with a brutal, pro-U.S. dictatorship led by the Shah.

This background is routinely omitted from mainstream public debate, making it harder for the public to see the current U.S.

intervention in Iraq as part of a larger historical pattern of the U.S. quest to control Middle Eastern oil. (Also left out of public debate is the role that U.S. interventions have inadvertently played in sparking the rise of a deeply anti-American Islamic fundamentalist movement in that part of the world.)

The possibility of an oil motive in the Iraq war is often dismissed on the grounds that it would be cheaper to simply buy the oil. Why bother invading, when you can get all the oil you want by just writing a cheque?

But the cheque-book solution was also available—and rejected by Washington—when Iran nationalized its oil a half-century ago. Tehran had every intention of continuing to sell oil to the U.S., and indeed was offering it at a very low price. Americans could have simply bought the nationalized oil.

But Washington wasn't interested in a cheque-book solution. The issue for Washington wasn't access to oil, it was control of oil. Without control, there is no guarantee of access. This point was driven home dramatically in 1973, when the Arab countries slashed their oil exports to punish Washington for purely political reasons. Thus, no cheque-book solution was even available. That Arab oil embargo left an indelible impression on U.S. strategic planners, who have since focused on ensuring America is never vulnerable like that again.

So spending $270 billion invading and occupying Iraq wasn't a calculated plan to get oil at a good price. Rather it seems to have been, at least in part, a calculated plan to secure Washington's crucial control over oil.

Although part of a larger historical pattern, the invasion of Iraq is easily the most extreme case of U.S. intervention over oil, and it may be a precursor of what lies ahead as we enter a far more precarious energy future with oil reserves in decline.

The simple truth is that oil is a finite resource. It is stored solar energy from the distant past—a one-time inheritance that we've foolishly squandered, consuming it recklessly for the past hundred years. At this point, there is not enough oil left in the earth's crust for us to go on living as we have for much longer.

One can debate exactly when world oil production will reach its peak and start to decline, and whether the crunch of declining supplies will start to affect our lives in the near future or not for another decade or so, but such a crunch is coming. The surging oil prices of the past year are just a small taste of the ramped-up competition for oil that lies ahead, and the far-reaching implications it will have for our lives. A report in February 2005 by the U.S. Department of Energy's National Energy Laboratory gives us a sense of the scope of the dilemma. "The world has never faced a problem like this," the report noted, ". . . Previous transitions (wood to coal and coal to oil) were gradual and evolutionary; oil peaking will be abrupt and revolutionary."

One suspects it may also be violent. The geopolitical struggle over oil—a rivalry that marked much of the last century—seems likely to intensify in the coming years. With the U.S. showing little interest in reducing its enormous consumption even as its domestic reserves are depleted, Washington is increasingly focused on securing control of reserves outside its borders.

Meanwhile, its chief rival, China, is experiencing explosive growth in its oil demand and is hungrily scouring the world looking for fresh sources of supply. China's increasingly aggressive search, including a bold bid in June 2005 to take over U.S. oil giant Unocal, provoked concern and even hostility in U.S. corridors of power. All this suggests that the world's most heavily armed nation and the world's most populous nation may soon find themselves facing off in a bitter struggle to lock up control of the world's most valuable resource—just as global production of that resource begins to decline.

And none of this even touches on the most devastating of all problems related to our oil addiction: global warming. So we're left in a strangely paradoxical situation: there's not enough oil to meet the world's growing consumption, but that growing consumption is itself threatening to imperil the world.

There are no quick fixes to the problem. Ultimately, it's going to involve changing the way we live in some fundamental way. But acknowledging the scope of our addiction would be a good beginning. Continued obfuscation over Iraq isn't helpful. A little more skepticism about invaders who come in the guise of liberators might help shine some badly needed light on one of the world's most combustible problems.

L.M.,
Toronto,
April 2006

CHAPTER 1

FORT KNOX GUARDED BY A CHIHUAHUA

"You give us the money. We give you the truck.
Nobody gets hurt."
—ADVERTISEMENT FOR THE 2003 HUMMER SUV

The Hummer SUV pictured along with this snappy ad copy is a massive fortress-like vehicle—something suitable for, say, taking the whole family for a spin through downtown Baghdad. With the playful, whimsical look of a Brink's truck, the Hummer practically sings out: "Out of my way, motherfucker." There's no mistaking you'll feel safe inside. But, of course, the joke in the ad turns on the old bank robbery line: if everybody just co-operates, "nobody gets hurt." That's where the fine line between sassy advertising copy and outright lying is crossed. In fact, huge gas-guzzling SUVs like the Hummer are one of the fastest-growing causes of global warming, with its potentially catastrophic impacts for human life on the planet.

Here then is how the ad should read: "You give us the money. We give you the truck. *Everybody* gets hurt."

———

If nothing else, Washington's saber-rattling against Iran in the spring of 2005 should have evoked a sense of *déjà vu*. The Middle Eastern country was said to be run by very bad men who oppress their own citizens and who are determined to acquire weapons of mass destruction. Even the name of the country is strikingly similar: *Iran*, so like *Iraq*. Both roll off the tongue with ease, both are well endowed with the world's most valuable commodity, and both conjure up frightening images of men who prefer weird Biblical outfits to proper business attire. So it turned out to be an easy transition. Without a blush of awkwardness, media commentators began preparing the public for a new reality: Washington might have to intervene in *Iran* in order to protect the American people and bring peace to the world.

If so, it wouldn't be the first time. Washington intervened in Iran back in 1953, after Iran nationalized its oil industry. The U.S. orchestrated a coup that overthrew Iran's democratically elected government, and installed a pro-U.S. dictatorship. This led to the rise of a fiercely anti-American Islamic fundamentalist movement that eventually took control of Iran and spread throughout the Middle East and beyond.

This background rarely makes its way into the current debate over Iran, nor is consideration given to the possibility that Washington might be motivated in part by a desire to regain control over Iran—along with its considerable oil reserves.

Instead, the media keeps its focus on Washington's allegations that Iran plans to develop nuclear weapons—just as it kept its focus on Washington's allegations that Iraq had weapons of mass destruction.

(Ironically, it was Washington that first supplied Iran with nuclear technology back in the days when it was a U.S. ally. And it has been Washington's unceasing hostility to Iran's Islamic revolution that has encouraged the country's Islamic rulers to think of developing a nuclear deterrent. The International Atomic

Energy Agency has not, however, found Iran to be in contravention of the Nuclear Non-proliferation Treaty.)

The media learned a bitter lesson when CBS anchor Dan Rather blindly trusted the credibility of a source discrediting President George W. Bush. But as for trusting the Bush administration—which has already gone to war over weapons that didn't exist—no lesson has apparently been learned.

———

Meanwhile, Washington's ongoing intervention in *Iraq* was now said to be on the right track.

With the turn-out of millions of Iraqis in the elections of January 2005, there was a giddiness among Washington war planners not seen since jubilant Iraqis had toppled the statue of Saddam Hussein, with the help of an American tank that happened to be on hand. After almost two years of unrelentingly bad images from the Iraqi war front, here finally were some good-news images to feast on—Iraqis dancing with joy in the streets, celebrating their dramatic experience with a ballot box. For supporters of the U.S. invasion, long pushed onto the defensive, this was indeed a moment to savour, a moment to celebrate how justified the invasion had been all along. This point was made repeatedly. Anyone who had opposed the war was now pretty much exposed as an enemy of democracy, and as an unrepentant Saddam-lover.

To listen to the giddy media commentary, one could easily have concluded that Iraqis had voted to show their support for America. Yet, the one platform common to all parties that took part in the Iraqi election had been the need to end the U.S. occupation. "Many of the voters came out to cast their ballots in the belief that it was the only way to regain enough sovereignty to get American troops back out of their country," noted Juan Cole, a professor of Modern Middle Eastern History at the University of Michigan. Some voters may have had simpler motivations; there were reports that proof of voting was necessary for access to food rations. Certainly it was a stretch to interpret the electoral results as encouraging for America. Ibrahim al-Jaafari, who emerged as prime minister after several months of post-election wrangling, was affiliated with the Dawa Party, a fiercely anti-American group believed to be implicated in the 1983 bombing of the U.S. embassy in Kuwait.

Besides, if there was a "hero" of the emerging Iraqi democratic process, it wasn't U.S. President George W. Bush, but rather Grand Ayatollah Ali al-Sistani, the moderate, influential Shi'ite leader who had doggedly pushed for elections right from the start, and sent hundreds of thousands of his supporters out onto the streets to back up his demands. Bush, on the other hand, had doggedly resisted elections right from the start, preferring that Iraq be run by a U.S. proconsul, with a new constitution to be drawn up by a few hand-picked exiles. "If it had been up to Bush, Iraq would have been a soft dictatorship," according to Cole.

When Washington finally agreed to elections, Sistani, seeing an opportunity for his long-oppressed Shi'ites to gain political clout, issued a *fatwa* making it a religious duty to vote. So chalk up the big turn-out on election day to enthusiasm for democracy—and loyalty to the ayatollah.

Next time Bush wants to "liberate" a country, we'll no doubt be shown post-election footage of Iraqis dancing in the streets, without any acknowledgement that those joyous Iraqis were probably celebrating the first step in pushing foreign occupiers out of their land.

The media's portrayal of the Iraqi elections as a triumph of democracy was yet another step in the ongoing presentation of U.S. actions in Iraq as a tale of good intentions. Keeping to this narrative has been challenging at times—particularly back in March 2003, when Washington launched its invasion.

The invasion brought to an abrupt end the United Nations weapons inspection that had been proceeding methodically for months. Suddenly, the whole orderly process had to be forcibly shut down so that Washington could begin dropping bombs on Baghdad, a city of five million people—an attack that U.S. Defense secretary Donald Rumsfeld promised would be "of a force and scope and scale that is beyond what has been seen before." Without a trace of irony, George W. Bush had explained in a televised address that he was dropping these bombs "to make the world more peaceful." (One can only imagine what he might do if he were trying to make the world more violent.) So, instead of nightly footage of Iraq destroying its Al-Samoud missiles under the watchful eye of the UN inspectors, our TV screens were suddenly filled with images of explosions and buildings burning in Baghdad.

But these seemingly hostile actions somehow appeared rather benign on U.S. television, which covered the war in a curiously upbeat manner. Every TV station had its own in-house military experts, equipped with coloured pens to trace troop movements—like weathermen showing an approaching cold front or sportscasters sketching a particularly good play in the backfield. And every station had its own war logo ("Target Iraq," "Attack on Iraq," "Strike against Iraq.") A more appropriate logo for CNN would have been: "The Joy of War" or "Kicking Ass." With a CNN reporter describing an American tank rushing towards Baghdad as "the most lethal killing machine on earth," anchor Aaron Brown could hardly conceal his excitement. "Are you dazzled by what you see?" he asked, turning to CNN in-house general (and later Democratic presidential candidate) Wesley Clark. Together the two men marvelled at the American killing machines speeding across the sand.

As a massive phalanx of U.S. troops moved into Iraq, Rumsfeld publicly warned Iraqis that setting oil fields on fire would be punished as a "war crime." Clearly, it's one thing to drop mega-bombs on a densely populated city, quite another to do something really evil—like destroy a perfectly good oil well.

Rumsfeld's comment might have been seen as a clue that oil was a key concern of those who had ordered the invasion. But such a notion was vehemently denied, including by Rumsfeld himself, who declared: "An Iraq war has absolutely nothing to do with oil." The denial was also, curiously, insisted upon by most commentators in the mainstream media, who were quick to roll their eyes at the very suggestion.

The media barely mentioned the fact that Iraq was bountifully endowed with oil, ranking second only to Saudi Arabia in terms of the sheer size of its reserves. But, unlike Saudi Arabia, Iraq's oil is largely untouched. This makes it uniquely tantalizing to the major international oil companies, which in recent years have been desperately searching for new reserves to develop. "There's no question all these companies are licking their chops waiting for an opportunity to jump in," notes Fadel Gheit, senior oil analyst at the Wall Street firm Oppenheimer & Co. As Washington prepared for war, there was almost none of this sort of commentary in the mainstream media.

Indeed, as the Iraq saga unfolded in the months that followed—moving seamlessly from a tale of the disarming of a dangerous dictator to a battle to bring democracy to the Middle East—oil remained strangely offstage, hidden in plain sight.

—————

The future of Hugo Chavez, president of oil-rich Venezuela, started to look significantly more precarious in January 2005. It was then that announcers on CNN began referring to Chavez as a "Latin American strongman."

CNN announcers wield no actual power. But they are a barometer of accepted political wisdom in Washington, of which notions can be stated as fact without causing controversy. Calling Chavez a "strongman" is a case in point. In the popular parlance, the word suggests a dictator, someone who maintains power through force, not through the ballot box. So when CNN announcer Kitty Pilgrim referred to Chavez as a Latin American "strongman," she was positioning him in a long line of repressive South American dictators. This is a significant distortion. In fact, Chavez's democratic credentials are impressive; he's won two nationwide Venezuelan elections, and in the summer of 2004 he handily won a national referendum on his leadership—a referendum that was given the stamp of approval by the Carter Center, the human rights organization established by former U.S. president Jimmy Carter.

But the "strongman" label went unchallenged. To millions of CNN viewers he was presented as simply a dictator, a mini-Saddam. And as such, he is unlikely to be missed by those viewers if they learn at some point he's been overthrown—a fate that Chavez is convinced the Bush administration has in mind for him. This isn't such a far-fetched notion. Chavez was briefly overthrown in a coup in April 2002, and Washington was at least supportive of the coup leaders, if not actively involved. And if the Bush administration had been interested in ridding themselves of Chavez back in 2002, it is even keener to see him ousted now. Chavez has long offended Washington with his close relations with Cuba, his open defiance of American hegemony in Latin America and his redirecting of Venezuelan oil revenues from American oil companies to Venezuelan social programs. But in December 2004, the flamboyant Venezuelan leader really stepped over the line: he made some far-reaching deals with China to develop Venezuela's considerable oil reserves.

Oil is the lifeblood of the global economy. Certainly any country wanting to maintain its superpower status must be assured of access to oil—the fuel that for the past century (and for the foreseeable future) has been indispensable for both economic growth and military power, as well as essential to support the self-indulgent western lifestyle. So oil is clearly vital to the United States. But there's a problem: while America is the biggest consumer of oil in the world, it has relatively little oil of its own, and what it has is rapidly dwindling.

While Americans consume roughly 25 percent of all the oil produced in the world each year, the U.S. has only 3 percent of the world's oil reserves. There's an enormous gap, then, between what Americans use (and what their government intends for them to keep on using), and what they need (or are determined to have). This leaves the U.S. highly dependent on foreign oil imports—a dependency which grows with each passing year. The U.S. now imports more than half its oil; by 2020, it is expected to import more than 65 percent. This makes America vulnerable, in danger of running short of the commodity it most needs to remain the world's dominant superpower. And vulnerability is not something Washington accepts lightly.

In America's quest for global dominance, no power looms more ominously on the horizon than China, with its tumultuous economic growth and massive scale. That growth is made possible by China's voracious and ever-rising consumption of energy. In the last few years, China has accounted for roughly 40 percent of the growth in global oil demand; it now ranks second in the world in oil consumption, after the U.S. And, like the U.S., China is highly dependent on foreign oil. With the oil consumption of these two superpowers continuing to grow rapidly, an increasingly fierce competition is shaping up between them over the world's most valuable resource. And then there's India—another country with an enormous population and an exploding appetite for energy.

Only one thing needs to be added to fill out this thumbnail sketch of the coming struggle over oil: the world is much closer to running out of oil than most government or industry officials are willing to admit—a subject we'll return to shortly. The earth's dwindling reserves will inevitably make the competition to gain control of the world's oil all the fiercer. As Edmonton-based energy economist Mark Anielski bluntly puts it: "There's not enough oil to feed two superpowers."

Which brings us back to Chavez. Venezuela has the largest oil reserves outside the Middle East, and is one of America's chief suppliers. Until Chavez's rise to power in the late 1990s, Venezuela's oil was largely under the control of U.S. oil companies and the U.S. government. Venezuela has traditionally been regarded in Washington as a relatively stable and secure source of oil, compared to the more volatile Middle Eastern sources. But that close U.S.-Venezuelan relationship has changed under Chavez, and his recent deals with China signalled a particularly bold—and unwelcome—independence on the oil front.

Chavez had made no secret of his desire to look beyond the U.S. for markets for his oil, and the Chinese quickly stepped forward to offer themselves as customers. The co-operation agreements signed by China and Venezuela reflect China's keen interest in developing a close long-term relationship with the world's fifth-largest oil exporter. Many of the agreements actually involve marginal, declining oilfields, but they also open the door to China taking a future role in developing new areas with enormous potential, such as the Orinoco oil belt in central Venezuela and offshore natural gas deposits.

In fact, China is only one of Chavez's controversial new energy partners. Perhaps even more provocative from Washington's viewpoint, Chavez signed a series of oil and gas co-operation deals in March 2005 with Iran, a full-fledged member of Bush's "Axis of Evil." After meeting with Iranian President Mohammad Khatami in Caracas, Chavez waded right into the conflict between Iran and the U.S., openly siding with Iran. The flambouyant Venezuelan leader insisted that Iran "has every right" to develop a nuclear energy program, despite U.S. objections that Iran's interest in nuclear energy is merely a cover for its intention to develop nuclear weapons. Chavez also offered to support Iran should it become the target of U.S. aggression. "Before the threats of the government of the United States against the brother country of Iran, the Iranians can count on our support, our affection, and our solidarity."

All this has contributed to the antagonism between Chavez and the White House, which has intensified since Bush's re-election. The administration has stepped up its public criticism of Chavez, accusing him of supporting leftist guerillas in Colombia and of suppressing press freedom at home (even though the fiercely anti-Chavez media operates freely in Venezuela). The White House has also urged Venezuela's neighbours to distance themselves from him, and is pondering measures to distance itself further from his regime, including greater support for Venezuelan opposition forces—the same forces that overthrew Chavez in 2002. The administration's animosity was perhaps most palpable when Secretary of State Condoleezza Rice, during her Senate confirmation hearings, mentioned that she had "nothing good to say" about the Venezuelan leader.

Only a month after signing the oil deals with China, Chavez appeared at the World Social Forum, the worldwide gathering of social justice activists held every year in Porto Alegre, Brazil as a counterpoint to the annual celebration of capitalism at the World Economic Forum in Geneva. The sprawling crowd of activists in Porto Alegre was just the sort of event Chavez warms to. He loves the stage; he knows how to pump up a crowd. In Venezuela, he can hold an audience enthralled, speaking passionately and without notes for hours about the need to strip power from corporations, to redistribute the oil wealth, and to break Washington's stranglehold over Latin America. In his exuberance, he often breaks out singing, urged on by requests from the audience for popular love songs. He readily obliges, prompting the crowd to call for more. A Chavez performance is part speech, part song, part love affair.

To the 15,000 activists in Porto Alegre, Chavez is probably the closest thing to a hero, an increasingly loud and aggressive champion of social justice on the international stage. His reforms in Venezuela have been radical and far-reaching, and his willingness to take on Washington and the corporate world has been unusual, to say the least, among world leaders. The oil deals with China and Iran are just further evidence that Chavez is not afraid to go his own way, even at the risk of provoking the Bush administration. As he took to the stage in a packed sports stadium in Porto Alegre, the crowd went wild, greeting him like a rock star and chanting "Here comes the boss!"

So there he was: *the boss*, sitting on top of one of the world's biggest oil reserves, thumbing his nose at the U.S—all right in America's backyard.

Maybe to some he's the "the boss." But in the eyes of Washington and millions of Americans who watch CNN, Chavez is nothing more than "a Latin American strongman." And the world can surely do with one less of those.

———

If Chavez exudes feistiness, Stephen Harper could be said to manifest blandness. The Canadian prime minister can be counted on never to break out in popular song while delivering a political address, nor drive a stadium full of activists into a frenzy with his political passion.

Harper's blandness is a welcome relief in Washington, where Canada has long enjoyed the status of a hopelessly dull, but enormously co-operative neighbour—a neighbour who happens to be rich in oil, as well as natural gas and electricity. Canada is in fact the mirror opposite of the U.S.—generously endowed with energy resources, but sparsely populated. What a perfect fit, then, for the U.S., with its increasingly meagre energy supplies and its substantial population. And the two countries fit well in another way; while America is aggressive and determined to get what it wants, Canada is generally submissive and pliant, or at least eager to accommodate the desires of its big, brash neighbour. Locking up access to Canada's ample energy endowment has long been in the minds of U.S. strategic planners. And the ever-friendly Canadians have been politely obliging.

Today Canada is the largest single source of energy imports to the U.S. And its role feeding the U.S. energy appetite is growing. A three-volume report prepared by an influential Washington think tank, the Center for Strategic and International Studies (CSIS), with the help of a bipartisan Congressional team, highlights just how important Canada is in the U.S. energy picture of the future. "Canada has been quietly but gradually assuming a growing role as a major supplier to the United States of both oil and natural gas," says the report, *The Geopolitics of Energy into the 21ˢᵗ Century*. It notes that imported oil from Canada rose from 11.9 percent of total U.S. oil imports in the early 1990s to 14.3 percent by the end of the '90s. Meanwhile, U.S. dependence on Canadian natural gas "approaches 95 percent." The report observes that, when oil and gas are both considered, "Canada becomes the single largest provider of energy to the United States, exceeding Saudi Arabia, Venezuela or Mexico" and that "Canada is poised to expand sharply its exports of oil to the United States in the coming years."

Fine—as long as Canada doesn't want to change its mind about any of this. In fact, Canada *can't* change its mind about any of this—a point celebrated in the CSIS report. When Canada signed the North American Free Trade Agreement (NAFTA) in 1993, it gave up its right to cut back the amount of energy it exports to the U.S., unless Canada cuts back its own consumption by the same amount. (Interestingly, Mexico, which is also a party to the NAFTA agreement and is also well endowed with oil, refused to agree to this provision and was granted an exception.)

Canada's willingness to sign on to the provision raises an interesting question. What if there were an oil shortage—something not inconceivable in a future sure to be marked by energy uncertainty? The CSIS report raises this very scenario: "Would [Canadian] exports to the United States be redirected to Canadian consumers?" it asks. It then promptly answers: "No, because NAFTA requires that any shortages would be shared between the two countries." In other words, Canada wouldn't be allowed to reduce its exports to the U.S. (in order to redirect those barrels to Canada). Redirecting Canadian oil to Canadians isn't permitted, *regardless of how great the Canadian need may be*. Canadians could be left freezing in the dark while Canadian oil keeps on servicing Americans—a point long raised by Canadian critics of NAFTA but rarely so plainly acknowledged by Americans. It's a situation that seems to comfort the U.S. senators, congressmen and think tank analysts who wrote the CSIS report. With obvious satisfaction, they conclude: "There can be no more secure supplier to the United States than Canada."

The reliability of Canada is likely to become even more important, since Canada is one of the few countries in the world whose energy supplies seem about to get bigger, possibly much bigger. That's because Canada has a potentially enormous store of untapped oil in the "tar sands" of northern Alberta. Getting the oil out of the tar is a horrendous task; it involves a massive, high-tech operation that causes serious environmental damage and requires copious amounts of two precious resources: water and natural gas. The gas would likely be brought in by a proposed pipeline from the Mackenzie Delta near the Beaufort Sea—a pipeline that is fraught with its own environmental problems and has provoked resistance from environmentalists and some Native activists. Since natural gas causes less damage to the environment and fewer greenhouse gas emissions than oil, the tar sands raises the prospect that vast amounts of a relatively clean fuel (gas) will be used to produce a much dirtier one (oil). By any logic, then, most of that tar sands oil should be left in the ground. But, given the world's galloping energy desires, it certainly won't be. The tar sands are already producing roughly a million barrels a day, and will likely produce a lot more, as world oil prices move higher, making the high cost of extracting the oil from the tar less and less of an obstacle.

The prospect of this gigantic new reserve of oil—potentially as big Saudi Arabia's, although not nearly as accessible—has made Canada suddenly loom large on the U.S. energy radar. It also means that NAFTA may not go far enough, from Washington's point of view, in locking up U.S. access to Canada's energy. NAFTA prevents Canada from cutting back its exports to the U.S. But what if vast new amounts of Canadian oil from the tar sands were to become available? There is nothing in NAFTA requiring that that new supply go exclusively to the U.S. Once again, China and its voracious and growing energy needs come into view.

China is indeed eyeing the stretch of oil-soaked black muck stretching across a vast swath of northern Alberta. And in January 2005, even as he was in the midst of signing deals with Chavez to develop Venezuela's oil, Chinese premier Wen Jiabao entertained then Canadian Prime Minister Paul Martin in Beijing, where the two leaders signed energy co-operation deals. Most notable was the agreement to begin plans for a 720-mile oil pipeline from northern Alberta to the west coast of British Columbia, enabling oil from the tar sands to be shipped to China. The move did not go unnoticed in U.S. energy circles.

The possibility that a sizeable amount of a vast new oil supply from Canada would end up fueling the cars and factories of rival China is not at all what Washington and the U.S. oil industry had in mind. The need to keep Alberta's tar sands—or "oil sands" as they are more optimistically called—within the U.S. sphere of influence is stressed by an investment journal produced by the Chicago-based financial management firm Harris Nesbitt: "Alberta's oil sands must be developed as rapidly as possible, and *they must be developed by organizations friendly to the U.S. . . .* The U.S. majors, BP and Shell, should get much more deeply involved in oil sands development. There is no comparable opportunity anywhere in the world, *and ownership of those vast reserves that will take a century to deplete should be in reliable hands, and directed to pipelines that are directed to the U.S.* [italics added]"

This kind of thinking lies behind the recently stepped-up push by U.S. business interests to further lock up access to Canada's energy. Those plans are also aggressively being advanced by Canada's own business elite—much of which consists of local managers who ultimately take their orders from American corporate bosses. The business elites in both countries, and Mexico, have teamed up in a task force, with the support of all three governments, to consider plans for further integrating the economies of the three countries, possibly even into a customs union. Achieving a deal on "North American energy security" is a key focus of the task force. Since Canada and Mexico already essentially enjoy energy security (at least as long as the world's energy supplies hold out), the goal here is clearly to ensure that Canadian and Mexican energy is fully available to the U.S., so that it too will have energy security—something it currently lacks. In the process, however, Canada could risk losing its own energy security; indeed, it may have already compromised it. As energy researcher Julian Darley has noted: "The list of energy independent 'developed' countries extends exactly to three: Canada, Britain and Norway. Britain is about to lose that status, and Canada is not that far behind, thanks to NAFTA."

Meanwhile, the prospect of cozy energy relations between Canada and China is touching a raw nerve among some

Americans who are already angry at Canada for refusing to join in the U.S. invasion of Iraq and to support U.S. plans for a missile defence system. "Backstabbed!" was the provocative opening salvo of a newsletter sent out in March 2005 by an investment service known as the Daily Reckoning (run by the authors of the popular business bestseller *Financial Reckoning Day*). "The Canadian government is selling the United States' oil future down the river," declared the electronic newsletter, describing the proposed pipeline from northern Alberta to the B.C. coast as the "Great Canadian Doublecross." It continued: "Hear that Giant Sucking Sound? That's the Sound of U.S. Oil Supplies Being Siphoned Away by the Chinese." (Note that Canada's oil is described here as "U.S. oil supplies.") Hinting darkly of a growing anger against unfaithful Canada, the newsletter speculates about some of the feelings on Capitol Hill: "Those backstabbing Canucks do nothing but leech off our economy. . . . They owe us that oil! . . . They spend next to nothing on defense because of their lucky proximity to our awesome military might. . . . Without our protection, Canada is the natural resources equivalent of Fort Knox guarded by a 'No Trespassing' sign and a Chihuahua. Can't we just take the oil?"

———

In the contentious debate over global warming, one hesitates to suggest that Armageddon is just around the corner, that it could arrive even before Larry King's next divorce. Nobody wants to leave oneself vulnerable to the charge of being a Chicken Little alarmist—a charge routinely thrown around by the small holdout group of global warming skeptics.

So it's interesting to note that the Chicken Littles have already infiltrated the Pentagon—one of the few places that, along with the American Petroleum Institute and the White House, seemed impervious to fears about where our excessive oil consumption is leading. No longer. A report commissioned by influential Pentagon adviser Andrew Marshall and presented to the Pentagon in October 2003, describes how global warming could transform the world dramatically in the next twenty years. It warns that major European cities could be sunk beneath rising seas and Britain plunged into a Siberian climate while other parts of the world—notably parts of the east coast of North America—could be forced to endure unusually cold, brutal winter storms as well as prolonged drought. Meanwhile, elsewhere in the world, typhoons, mega-droughts and famine could be expected. The report goes on to suggest that these catastrophic changes would likely lead to widespread human strife, rioting and even nuclear conflict. "Humanity would revert to its norm of constant battles for diminishing resources," the Pentagon analysis notes. "Once again, warfare would define human life."

One of the striking things about the Pentagon report—apart from its pedigree as a Pentagon document—is that it paints a scarier picture than reports by environmental groups generally do, perhaps because, after years in the trenches of this debate, environmentalists have learned to protect their credibility by not sounding overly alarmist. Imagine Greenpeace making claims that global warming could lead to nuclear war! But just as Richard Nixon could get away with making peace with China (because no one would ever accuse him of being a Communist), so the Pentagon can get away with setting out a truly hair-raising scenario about the impact of global warming (because no one would ever accuse it of harbouring sentimental feelings about the environment). And so the report pulls no punches.

Anyone who has contentedly stayed detached from the global warming scare stories—thinking that a solution will be found and that in the meantime a little more warmth on the beach wouldn't be such a bad thing—would get a shock from reading this partic- ular Pentagon paper. Perhaps the most immediate adjustment required to the benign view is the notion that what we're facing is the prospect of *warming*. To a northern sensibility, that might not sound so bad. Melting polar ice caps, while undesirable for polar bears, hardly sound like particularly bad news for the rest of us. But, as the report points out, recently gathered evidence suggests one likely impact: the dysfunction of the Gulf Stream. In the great interconnected world weather system, the extra quantities of cold fresh water created by the melting polar ice caps and glaciers could effectively prevent the Gulf Stream from doing what it does so well—bringing warm water from the southern Atlantic and, in the process, sparing Europe and eastern North America from the grim, sub-Arctic weather that populations in those regions would otherwise be obliged to endure. (When this same system broke down 12,700 years ago, the report notes, there were icebergs off the coast of what is now Portugal.) If "global warming" sounds benign and even comforting, like the warm embrace of the sun on an early spring day, how does "global *freezing*" sound?

Another comforting thought has been the notion that human ingenuity will bail us out. The same kind of brainpower that thought up the internet and the flat-screen TV can surely (if the incentives are generous enough) figure out a solution to this too before things get out of hand. But things might already be out of hand. The report notes that there's been a tendency to regard global warming as a problem that will set in gradually, giving the world a chance to adapt and even possibly take advantage of what could be longer growing seasons. "This view of climate change may be a dangerous act of self-deception, as increasingly we are [already] facing weather-related disasters," the report states. "Rather than decades or even centuries of gradual warming, recent evidence suggests the possibility that a more dire climate scenario may actually be unfolding." Hence, the report's focus on "an abrupt climate change scenario" that has already begun.

Rather than gradual global warming, then, how about *sudden global freezing, starting tomorrow?*

The real point of the report, however, isn't just to draw attention to these enormously unpleasant scenarios, but to suggest that they should be treated as national security concerns. The extreme climate change scenario outlined is "plausible," according to the report, and "would challenge United States national security in ways that should be considered immediately." Indeed, the prospect of life on earth reverting to a primitive, desperate, brutal quest for survival suggests a terrorized form of existence quite as frightening as airplanes being seized or bridges being blown up.

The report—prepared for the Pentagon by business consultant Doug Randall and Peter Schwartz, former head of planning at Royal Dutch/Shell Group—was never intended for public consumption. But even after a copy was leaked to the prestigious London *Observer* in February 2004, the report attracted surprisingly little attention in the mainstream media. While virtually every conscious North American was aware that Janet Jackson's breast had been exposed on TV a few weeks earlier, only a tiny fraction of that public had been exposed to something that might be considered even more shocking: that our wanton over-consumption of oil might be about to create a whole new kind of terror in our lives. Yet the Bush administration, which had consistently ignored and downplayed the threat of climate change and done its best to sabotage the international Kyoto accord aimed at dealing with the problem, was not about to change horses in its "war on terror." Its defence strategy would remain fixated on shadowy men in long-flowing robes, not on ones wearing business suits and bearing large cheques made out to the Republican Party.

A lot of people seem to think that humans are so smart we'll figure a way out of our energy dilemma. Maybe so, but the past is
hardly encouraging on this score. Certainly, the world's handling
of our energy resources, since their value came to be appreciated
more than a century ago, has not been impressive. Consider, for
instance, what we've done with oil—the most effective and flexible form of energy available to us, and the one that has powered
the industrial revolution. Oil is finite and not recyclable. There
is a fixed amount of it embedded in the earth's crust, and once
we've pumped it all out, there won't be any more. And yet, in the
twelve or so decades since we first started using oil in earnest,
we've already used up roughly half of all the oil that the planet
has to offer. And the first half is the easy stuff to reach. Soon it
gets harder.

This, then, is the second part of the energy dilemma, and it
speaks to an astonishing lack of foresight on the part of the human
race. We've used up the earth's oil so rapidly and recklessly that
we have not only jeopardized the viability of the planet (part one
of energy dilemma), but we have, at the same time, squandered
much of this incredibly valuable one-time inheritance.

This may sound like a contradiction. If oil is so bad for the
earth's ecosystem, why should we care if it's running out? The
problem is that we've built our lives around it. Virtually every
aspect of life in the modern world involves the use of oil in some
form. We rely on it for just about every thing we do, eat, dress,
type, watch, and move around in. Oil or derivatives of oil are
used to create our vast array of consumer goods and communication devices, as well as the fertilizer for much of the world's food.
And, of course, oil has become virtually essential to our mobility,
fueling our cars and our airplanes. Without it, life in the twenty-
first century would seem more like life in the 1800s.

Yes, we should move on to other less damaging energy sources.
But even an apparently promising solution like hydrogen—highly

touted as the energy source of the future—has serious unsolved problems. For one thing, it is almost always made from natural gas, which is a finite resource already in tight supply and much more difficult to transport than oil. Perhaps in time other more promising alternatives will emerge, but as of now, there is no energy source in view that is as effective, versatile, and potent as oil.

The real solution lies in us learning to get by on significantly less oil, as well as relying more on other, less damaging energy sources like wind, sun and waves. Ultimately, then, we're going to have to change the way we live. But we don't really want the modern world as we know it to grind to a sudden halt while we're still living in that world. Journalist George Monbiot aptly describes the corner we've painted ourselves into: "Either we lay hands on every available source of fossil fuel, in which case we fry the planet and civilization collapses, or we run out, and civilization collapses."

In the short run, civilization probably won't actually collapse. It will just get a lot more troubled. As we begin consuming the remaining half of the earth's oil, things are almost certainly going to get more problematic here on earth. Among other things, the scramble to get control of oil—a constant in the last century—is likely to get a lot more intense and perhaps more violent.

It's not that we're about to run out of oil. There are roughly one trillion barrels still in the earth's crust. At the rate we're consuming it, we have about another forty years' supply. By then, presumably, we will have figured out an alternative and adapted our world to it. But there's a more immediate problem long before the oil actually runs out; it reaches its *production peak*. After that, the amount of oil produced each year in the world starts to decline. That may not sound very ominous. But once this peak is reached—roughly half way through the extraction of the world's oil supply—the easily accessible oil is gone and getting out the remaining oil becomes considerably more difficult and expensive.

This notion that oil production has a "peak" was first conceived in 1956 by geophysicist M. King Hubbert, who worked at the Shell research lab in Houston. Using this concept, he predicted that oil production in the U.S. would peak about 1970—a notion that was scoffed at at the time. As it turned out, Hubbert was correct; U.S. oil production peaked in 1970, and has been declining ever since. Hubbert's once-radical notion is now generally accepted among geologists.

For the world as a whole, the peak is fast approaching. Colin Campbell, one of the world's leading oil geologists, believes that global oil production will reach its peak some time in the next few years, if the peak has not already been reached. "The world started using more oil than it found in 1981," Campbell said in a telephone interview from his home in Ireland. "We've been eating into that reserve of past discoveries ever since." Campbell's view, while rejected by many oil analysts, particularly economists, has gained increasing exposure and acceptance since he published an influential article on the subject in *Scientific American* in 1998. Kenneth Deffeyes, a geophysicist at Princeton and author of *Hubbert's Peak: the Impending World Oil Shortage*, has been another influential voice predicting a fast-approaching world peak.

Even optimistic scenarios of exotic oil development from Alberta's tar sands or beneath the depths of the ocean won't ultimately change the picture, according to Campbell. He insists that the notion that the tar sands could be the answer to the coming energy scarcity is highly misleading. "No one disputes the resource is enormous," he says. "But getting it out of the ground is a costly, slow process." So far, only the most easily reached parts of the tar sands have been tapped, and getting the rest out will prove difficult and slow, making little dent on the declining world oil supply situation, he maintains.

All this would be fine if the world's appetite for oil were declining in tandem. But the opposite is happening. Even as the discovery of new oil fields slows down, the world's consumption speeds up. For every new barrel of oil we find, we are consuming four already-discovered barrels, says Campbell. The arithmetic is not on our side, and promises to get worse. Consumption is expected to increase by at least two percent a year over the next few decades. As the Pentagon paper on abrupt climate change points out: "[G]lobal demand for oil will grow by 66 percent in the next 30 years, *but it's unclear where the supply will come from.* [italics added]"

Campbell's message is echoed, with equal urgency, by Matthew Simmons, a prominent Houston-based investment banker who has served as an energy advisor to George W. Bush. Simmons notes that most of the oil the world relies on is coming from oilfields that have been in production for more than thirty years and are much closer to exhaustion than is commonly realized. Meanwhile, he notes, "oil demand has become a runaway train." In the early 90s, most experts thought world oil demand was permanently stuck at about 66 million barrels of oil a day. Now, just slightly more than a decade later, that demand has increased by almost one-third, and is projected, thanks to the voracious energy appetites of the U.S. and China, to keep ratcheting upward.

Given the declining production in older oil fields and the rising demand, Simmons predicts the world will soon need to start producing another 8 million barrels of oil per day. "This is equivalent to a new Saudi Arabia." But there is no equivalent to Saudi Arabia. For that matter, Simmons maintains that Saudi oil reserves are more depleted than is appreciated. Simmons ends up sounding just as alarmist about the prospects of the world running out of oil as the analysts who wrote the Pentagon report were about dangers of global warming. "This problem ranks alongside thermonuclear war," Simmons insists, pointing out that, unless it is solved, among other things, the world will lack the energy to produce drinkable water and that transportation will revert back to "wind and foot power."

This would seem, by any meaningful standard, to be a problem worthy of serious attention at the very highest levels. But, oddly, it's a problem that is largely unacknowledged in official quarters. Instead, government and industry spokespeople exude a kind of confidence about the future that is tied up with a belief in the endless wonders of modern technology. The fact that oil is a finite resource, that there's only so much of it in the world, and that, once used, it's gone forever, is rarely allowed to darken the official story. In fact, this reality is actively denied. "The world's supply of oil is not finite," declared David Frum, former Bush speechwriter and now a fellow at the American Enterprise Institute in Washington. Frum's argument is that the market will simply correct any shortage of oil, that if the price goes high enough, more oil will be discovered. He compares oil to tomatoes; if there is more demand for them, the supermarket will stock more.

But oil is not the same as tomatoes, which produce seeds that enable us to grow more of them if we want to. Oil is the product of a specific set of climatic conditions that existed only once in the earth's 4.5 billion-year history. It is solar energy, trapped and stored in decaying plants during the Mesozoic Era, some 225 to 65 million years ago, when the earth was covered with swamps, huge trees, ferns and algae. The decomposing plant material took millions of years to develop into its present form. And while there's a great deal of oil in the ground, giving the impression of a limitless supply, in fact most of the earth's crude has already been discovered. "All the promising areas have been thoroughly explored," notes Colin Campbell. "There are good reasons why other areas receive little attention. There are of course details to fill in, but the general picture has become very clear. Some oil economists claim that if all the basins of the world were drilled as intensively as Texas, they would yield a huge amount of new oil. They are utterly mistaken for well understood scientific reasons." Overall, throughout the world, fewer and fewer oil fields are being discovered; the rate of discovery of new fields has been declining for about forty years. "There is only so much crude oil in the world," Campbell notes, "and the industry has found about 90 percent of it." The same can't be said of tomatoes.

The industry's failure to find new oil fields cannot be blamed on lack of trying. Huge resources and the most advanced technology have been directed towards the problem, yet even the major oil companies have been coming up short—a story that has received relatively little attention. To some extent, the sky-high profits the industry has enjoyed recently have masked the fact that Big Oil is having trouble coming up with new sources of supply. This is a serious problem for the industry, as well as pointing to a potentially serious problem for an energy-hungry world. A company's oil reserves represent its wealth, and its ability to earn income in the future. For a company to continue to prosper, it must keep expanding its reserves, or at least not let them shrink. The rule of thumb in the industry is that a company should at least replace as much oil as it produces each year. Yet, the industry has recently fallen considerably short of this 100 percent annual replacement rate.

The most dramatic example has been Royal Dutch/ Shell, the third largest oil company. In 2004, Shell reported a dismal replacement rate of only 15 to 25 percent. Early in 2004, Shell had shocked the oil world by admitting that it had significantly overstated the size of its reserves. In the following months, it continually had to revise its reserve estimates downward and, by early 2005, the company acknowledged that its reserves were about one-third smaller than it had been claiming.

Shell's trouble finding new sources of oil is in fact typical. "The entire industry is straining to replace its stores of energy," due to "industry-wide difficulties in finding oil," the *Wall Street Journal* noted in February 2005. Shell's problems have simply been more visible. This is because the other major oil companies have been buying up other companies and thereby acquiring *their* reserves. Since Shell hasn't got into the merger game, it has faced the reality of dwindling oil stocks sooner than the other giants. Wall Street oil analyst Fadel Gheit calls the industry's failure to find replacement reserves "alarming." Says Gheit: "For the first time in history, the seven largest companies have failed to replace their reserves." Matthew Simmons agrees there are serious problems with oil company reserves: "A significant amount of [their] stated proven reserves would never pass third-party . . . standards," he notes. All this suggests that we are farther along the road of oil depletion than the official numbers suggest.

It also points to higher oil prices. The investment bank Goldman Sachs sparked wide concern in March 2005 with a report predicting oil prices as high as $105 a barrel, as the world enters "the early stages of what we have referred to as a 'super spike' period," based on tight supplies. Such high prices would inevitably lead to reduced oil consumption, which would then presumably bring the price down. This has been the pattern in the past; the high oil prices of the 1970s and early 80s led to reduced consumption and then a significant price drop in the 80s. This comforting scenario may no longer apply, however. Jeff Rubin, chief economist at CIBC World Markets in Toronto, believes that things will be different in the future—due to the "accelerating depletion of conventional crude supply." Rubin notes that the earlier oil shocks of the 70s and early 80s were based on political events, as the oil-producing nations temporarily restricted supply. "This time around," he insists, "with suppliers already running full tilt, there's no tap that can be suddenly turned back on."

Even as it's struggling with dwindling supplies, the oil industry has contributed to the illusion of ever-bountiful oil, dismissing those who suggest we're running out of oil. It's interesting to note, however, that the industry was similarly dismissive in the 50s and 60s, when experts like M. King Hubbert pointed out that U.S. oil production would soon peak and then decline. For decades before that, the U.S. had been the world's biggest oil producer and was believed to have the world's largest reserves. Washington's failure to listen to warnings about the coming peak in U.S. production was to have devastating repercussions. Not only did the U.S. fail to take measures to conserve its reserves, it actually imposed quotas limiting the amount of foreign oil allowed into the country—quotas which resulted in U.S. reserves being depleted more quickly. We'll return to this tale of wilful ignorance and misguided policy in Chapter Three.

The foolishness of these import quotas wasn't fully appreciated until 1973, when the Arab oil embargo abruptly left the U.S. scrambling for all the oil it could get. Stunned by its vulnerability, Washington immediately dropped the quotas. But it was too late to avert the damage. So much U.S. oil was already gone, having been used up at an accelerated pace for fourteen years while the consumption of foreign oil had been limited. Washington had fast-forwarded itself towards oil dependency—a problem that would haunt it for years to come. And it had all been on the advice of the oil industry, with its reassuring insistence that there was plenty of oil—a reassurance the industry continues to give today.

———

Although now long forgotten, the advances taking place in Britain's textile trade in the first half of the eighteenth century were stunning—and very much on the minds of the out-of-work spinners who ransacked the house of Yorkshire woolen manufacturer John Kay in 1745. Kay had come up with a way to significantly improve the loom (the manually operated machine which lay at the centre of Britain's early rise as an industrial power). Up until Kay's brainwave, it had taken two weavers to throw the loom's wooden shuttle back and forth in order to create the cross-wise action necessary to weave yarn into cloth. Kay's idea was to mount the wooden shuttle on little wheels, and attach it to a spring, which could be set into action whenever the weaver tripped a cord. The ingenious little idea changed the nature of weaving; the awkward, laborious process of moving the shuttle repeatedly back and forth between two sets of hands was now history. With Kay's "flying shuttle," one weaver could do it all, and in about half the time.

This breakthrough meant a sudden drop in the demand for weavers, and it made the precariousness of the lower-ranked spinners—who fed the yarn to the more highly skilled weavers— even more acute. Now their services were in even less demand, and they knew exactly who was to blame as they set about trashing Kay's house and everything in it. Kay, prescient as always, could tell things were going to turn out badly as soon as the mob broke through his front door. Using the same cleverness that had led him to think up the flying shuttle, he managed to disguise himself sufficiently to pass as a member of the looting mob until he could slip unnoticed out the back door.

Kay's fate was not unusual for those who broke new ground and, in the process, risked causing great harm to the livelihoods of others. James Hargreaves, a chubby hand-loom weaver, also discovered this sad truth after a burst of creative thought one day in 1764 led him to figure out that the normally horizontal spindle of a loom would work more effectively if it were vertical—an idea that struck him while observing the spinning wheel that his wife Jenny had left overturned on the floor. Although he was initially careful to keep his improved loom within the family, the temptation to cash in on his technological advance led him into the business of selling a redesigned machine with eight vertical spindles, which, in honour of his wife, he called the spinning jenny. Its multiple upright spindles were best worked by the small, nimble fingers of children. Their sudden influx into the textile business represented a serious threat to large-fingered, grown-up spinners, who, in 1768, nimbly destroyed Hargreaves' original spinning jenny as well as everything else in his house.

That same year, a young barber's apprentice by the name of Richard Arkwright found himself drawn to the same compelling puzzle that attracted so many bright minds in the late 1760s—how to make an even better spinning machine. Arkwright figured out a way to improve the process by strengthening the thread used—an innovation that paved the way for the spinning of cotton cloth. He was also savvy enough to ensconce himself in a rented house surrounded by a thick growth of prickly gooseberry bushes. Still, when Arkwright noticed the local townspeople of Preston looking at him suspiciously, he moved to Nottingham, considered a safer town. There he designed a whole new large-scale spinning frame pulled by horses. He soon relocated again to a riverside spot, where horses could be replaced with water power, enabling the large-scale production of dazzling, brightly coloured cotton calicoes and ushering in, as well as the beginning of the modern factory, the era of the colourful social event known as the calico ball.

As the world went on to enjoy these innovations, the discarded workers of the textile trade could do little more than sulk, starve and, occasionally, demolish the offending machines. For this resistance they paid a heavy price. The authorities of the day enacted brutal punishments, including the death penalty, for those caught destroying the new machines. History has also not been kind in its characterization of these discarded workers, often called Luddites, who are almost always bad-mouthed as vengeful stick-in-the-muds, imprisoned by their own self-interest and unable to see the broader benefits for mankind in the changes they so fiercely resisted.

A parallel of sorts could be drawn between the Luddites and a very different sort of group fiercely resisting change today. Today's group doesn't go about breaking into people's houses or smashing new machinery that threatens them. They don't need to. They're running the world.

———

One would expect the inner suite of senior management offices at the world headquarters of ExxonMobil, the world's biggest corporation, to exude a sense of power. But one perhaps comes closest to getting the true feel of this inner sanctum in Irving, Texas from the term used by industry insiders to describe it—"the God pod." At the very top of the God pod for many years sat Lee R. Raymond, Exxon's chairman and chief executive. A cover story on Exxon in *Business Week* may have understated things a bit by referring to Raymond simply as "the Man."

The job of Exxon chairman entails many things, but ultimately, it boils down to this: keeping Exxon where it's been for the last century and a half—at the very top of the corporate world, the biggest, richest corporation on the planet. And that means the company must keep growing, managing a vast and complicated empire of oil and gas production, refining and distribution around the world. Of course, selling the world's most indispensable commodity has never been a bad business to be in—particularly for the small group of companies that straddle the top of this privileged world. But never more so than now.

Profits for Big Oil have been spectacular since the beginning of this decade and, with oil prices rising sharply since the spring of 2004, they just keep getting better. ExxonMobil has led the pack, reporting a record $36 billion profit for 2005—the biggest profit ever recorded by a corporation in history. And it's a particularly good time to be the most profitable company. While excessive corporate profits used to be a hot political issue, stirring up populist protests and prompting congressional investigations, there's little of that now. Ironically, in February 2006, Congressman Joe Barton of Texas launched an anti-trust investigation into an oil company. But his target wasn't Exxon. Rather, he was after Citgo, owned by the Venezuelan government, because it had offered discount prices on heating oil to poor communities in the U.S.

Even global warming, which is potentially the greatest threat humankind has ever faced, has proved to be a manageable problem—from Exxon's point of view. Since the scope of global warming came to be understood in the 1980s, one might have thought that the business prospects of Exxon and the other giant oil companies would dim. Surely the world was going to put serious effort into averting this rapidly approaching disaster. But it hasn't turned out that way. Despite virtually the entire worldwide scientific community urging action, little has been done to slow down the globe's ever-rising oil consumption. Indeed, consumption has risen at a sharper pace recently. That may be bad for the world, but it's good for Exxon. Lee Raymond played a particularly key role in making that happen.

As chairman of Exxon from 1993 to 2005, Raymond mobilized the vast resources of his company to make sure that the world remains hooked on oil. Perhaps more than any single individual, he fought to block the implementation of the Kyoto accord. Exxon's aggressive ten-year campaign culminated in George W. Bush's dramatic decision, two months after taking office, to withdraw U.S. support for Kyoto, seriously jeopardizing the world's chances of addressing global warming before the damage becomes irreversible. This, more than anything, is the true measure of Exxon's exceptional power—the power to block the world from taking action to save itself, because that action might hurt the company's stock prospects.

Yet Raymond has generally been portrayed in the media as simply being a smart, savvy corporate leader. The *Business Week* cover story presented him as open to "alternative energy sources," although skeptical that they will be "cost-competitive" in the near future. Of course, this is exactly what Exxon would like us to believe—that the alternatives may be a good idea, but they're not really viable.

In fact, alternative energy sources as well as cars that run on drastically reduced amounts of oil are possible, even already available on the market. What keeps these innovations from being fully "cost-competitive" are the massive government subsidies given to virtually every part of the business connected to the fossil fuel industry and the traditional oil-consuming car—a subject we'll explore later. In addition, a peculiar loophole in U.S. regulations gives preferential treatment to SUVs, the most fuel-inefficient vehicles of all. As a result, the huge growth in the SUV market in recent years has left North America's current fleet of vehicles with the lowest overall level of fuel-efficiency in more than twenty years. Without this perverse loophole and without the multi-billion-dollar annual subsidies, the new, more fuel-efficient cars would actually be quite competitive.

Oddly, these perverse policies are said to have something to do with "freedom." Raymond likes to portray the tougher fuel efficiency standards in Europe, for instance, as an infringement on freedom that Americans would find intolerable. As he told the *Financial Times* of London: "In Europe, you like to tell people what kind of cars they ought to use. Most Americans like to make that decision themselves—that's why they left [Europe]." But it's hard to see how driving fuel-*in*efficient cars is a source of meaningful freedom, or even an advantage for Americans.

So, it turns out Lee Raymond actually has a lot in common with the spinners who trashed John Kay's house in 1745. Both the Exxon executives and the spinners were threatened by the fact that time had moved on, that the world had grasped something about the future that made it difficult to simply continue as before. What's strikingly similar is the desire in both cases to hold back the clock, to cling to a status quo in which their place in the scheme of things was secured (at the very top of the corporate world, in Exxon's case, and somewhere above destitution, in the case of the spinners.)

Yet, while the spinners paid a heavy price at the time and have become the butt of long-running historical jokes, Lee Raymond enjoys staggering wealth and has long been celebrated in magazines as a forward-thinking business leader. If the comparison between the eighteenth-century Luddites and the Luddites of today seems harsh, one wonders if it isn't the spinners who've been unfairly maligned. It's true that they were violent at times. But, in fairness, their backs really were to the wall, their families really were at risk of going hungry. By contrast, Lee Raymond and other top Exxon officials are clearly not in any kind of desperate situation; they are part of the most privileged class of managers in history.

It's hard not to conclude then that, as Luddites go, this new twenty-first century batch is more offensive. After all, denying the world the opportunity to save the earth's ecosystem—which sustains nothing less than life on the planet—is, by any meaningful reckoning, a more grievous offense than denying the world the benefits of the flying shuttle or even the calico ball.

It's hard to make a mark in the highly competitive world of conservative Washington think tanks. The main ground of conservative thinking has been so exhaustively trod, the cause of freedom so repeatedly invoked, the evil of big government so thoroughly expounded upon. Newer think tanks on the scene are inevitably pushed further to the edges in order to be visible and thereby position themselves to claim some portion of the vast sums of corporate and foundation money on offer. At the up-and-coming Competitive Enterprise Institute—not to be confused with the long-established, more amply-funded and politically well-connected American Enterprise Institute—the volume is thus ratcheted up, the positions taken are a little wonkier. Kyoto is a big target; Kyoto defenders are dismissed as suffering from "an animus against humanity."

This phrase—"an animus against humanity"—slips effortlessly out of the mouth of Myron Ebell, the institute's director of global warming and international environmental policy. But what could it possibly mean? Ebell argues that environmentalists see humans as essentially bad. "[They believe] humans are evil, the use of human power is always bad; everything we do to nature is bad," says Ebell, a good-natured, smiling philosophy graduate, who has worked as a lobbyist for right-wing causes and politicians since settling in Washington in the early 1990s, after an upbringing in the rural American west.

Ebell, who confesses to not having an extensive science background, appears comfortable dismissing the scientific case about global warming, despite the fact that virtually the entire scientific community around the world has given it credence. "We think the case for scientific alarm looks weak," says Ebell, sitting in the institute's boardroom.

There is something almost funny about this blithe dismissal of findings supported by such a heavy contingent of scientists, including many Nobel laureates. Yet Ebell's Competitive Enterprise Institute, which has received more than $1 million from Exxon since 1998, is a voice that, remarkably, is taken seriously in the halls of power. (Officials in the Bush administration have even sought Ebell's opinion on tactics for downplaying claims about global warming, according to correspondence uncovered by the London *Observer*.) Ebell and the institute are part of a virulent industry-funded anti-Kyoto movement which has played a significant role in keeping Kyoto and other efforts to deal with climate change off the political agenda, both in Congress and in the White House. In the process, they have helped create an atmosphere in which science and world opinion are seen as suspect, as obstacles in the path of Americans living as they choose to live.

In fact, that's the way the issue has been presented to the American people—as a battle for America's right to do things as it pleases. The notion that there should be any curbs on the consumption of fossil fuels is presented to Americans as an infringement of some fundamental right. "Energy is fundamental to mobility, to comfort. When you start limiting people's access to energy, you limit their ability to live the way they want, to make choices," says Ebell. "We're opposed to things that limit people's choices."

Of course, Ebell conveniently overlooks the fact that the deterioration of the earth's ecosystem—and the resulting hurricanes, floods, tornadoes, droughts, insect infestations and crop destruction—also limits people's choices, and in much more serious ways. But beyond such practical considerations, there's another aspect to his argument that's more deeply disturbing. He implicitly rejects the notion that humans have a responsibility to each other, that living on earth is a shared experience, something done along with six billion other people (not to mention millions of other animal and plant species). Instead, living on earth is presented as an essentially individual experience, in which everyone simply maximizes his or her own mobility, comfort, pleasure . . . whatever.

This surely goes beyond any reasonable notion of individualism. Indeed, it makes a mockery of the American individualist tradition—one of the finest aspects of American democracy—and contorts it in the service of creating a monstrous sort of narcissism. One's own needs—or even just one's own whimsical desires ("I like the sporty look of an SUV")—take precedence above all else, including other people's most basic needs. The U.S. is responsible for roughly 25 percent of the worldwide emissions

that cause global warming, even though it has only 5 percent of
the world's population. Yet, if Americans want to drive bigger,
bulkier vehicles with less fuel-efficient engines, and that happens
to ruin the atmosphere for the whole world, well, man, get out of
their way! The notion that anyone could challenge this right,
that the six billion other people on earth might have a say in the
matter—or that their welfare should be considered — is rejected
out of hand.

What was that about an animus against humanity?

———

So rich in oil, so defenseless, so under the thumb of the demon-
strable villain Saddam Hussein, the miracle of Iraq is that it
didn't get brought firmly under U.S. control years earlier.

The amazing thing about the U.S. invasion of Iraq, when it
finally happened in March 2003, was not how quickly the feeble
nation succumbed to the onslaught of the mightiest military
force on earth, but rather how Washington had managed for
months beforehand to keep world attention riveted on Iraq's pur-
ported "weapons of mass destruction," rather than on the more
likely purpose of the conquest—oil.

To say such a thing, of course, is to put oneself at the margin
of public discourse. While protesters in rallies all over the world
that spring carried signs linking war to oil, no such crude linkage
was evident in mainstream commentary, which focused instead
on analyzing the legitimacy of the U.S. weapons case against Iraq
and recounting the evils of Saddam Hussein. There was certainly
plenty of skepticism about the war, even in sophisticated circles.
It was quite appropriate at a cocktail party, for instance, to question whether the UN couldn't do a better job of disarming Iraq
or whether the U.S. would have the stomach or patience to put
Iraq back together after the military victory had been completed.
These were serious questions, wrote ultra-sophisticate *New Yorker*

magazine editor David Remnick, who at the same time was dismissive of those who thought war was being driven by "a conspiracy of oil interests."

The conspiracy theorist is an object of derision, so accusing someone of being a conspiracy theorist is a sophisticated way of hitting below the belt. Conspiracy theories are widely seen as the territory of either the genuinely delusional or at the least the hopelessly naïve, those who fail to grasp nuance, who are only capable of seeing black and white when there are infinite shades of grey. A debate can often be brought abruptly to a halt when the charge of conspiracy theory is leveled. A senior analyst at the conservative Washington Institute thus easily quashed further probing in February 2003 when she brushed aside a question from a CBC radio interviewer about whether oil might be the real motive for the invasion of Iraq. "I think any rational analysis would expose that as a conspiracy theory," she said, closing the door firmly and conclusively on that line of questioning.

It's true that almost any question is infinitely complicated by dozens of factors. It's also true, however, that some factors are more important than others, and that, ultimately, people do things for reasons—reasons which are often self-serving and which they are loath to admit to. Choosing to overlook these self-serving motivations and to see complexities where simple explanations of greed or malfeasance suffice doesn't necessarily bring us closer to the truth—a point nicely captured in the cartoon of two cows dismissively laughing off alarmist theories about what goes on inside a slaughterhouse. Those clever cows understand that the people who run slaughterhouses are actually mostly interested in improving the well-being of animals.

Similarly, the sophisticates are able to see that the people who run the U.S. government are mostly interested in improving the well-being of the people of Iraq. This, at least, is what the sophisticates appear to believe, as they devote great attention to analyzing the alleged U.S. goals of turning Iraq into a model of democracy for the Middle East. So the sophisticates ponder questions like: Is democracy possible in the Middle East? Can a U.S. invasion bring it about? Will a democratic Iraq inspire other Middle Eastern countries to follow suit or just incite more Islamic fundamentalism? All these sorts of questions, which feed on each other, start from the basic assumption that the White House is motivated by a desire to bring about democracy in Iraq and elsewhere in the Middle East, even though there's not actually any evidence to support this contention (and quite a bit of evidence to contradict it, which we'll get to later). On the other hand, the notion that the White House might be motivated to take control of Iraq because Iraq has the world's second-largest reserves of a commodity that is key to controlling the global economy is dismissed as a simplistic conspiracy theory.

Of course, the Iraq invasion was not *just* about oil, and we'll return in the next chapter to a discussion of other U.S. motives in Iraq. Still, it was striking how mainstream pundits abandoned all skepticism in their willingness to accept Washington's assertion that it was not at all motivated by oil, that its real goals involved making the world safer and bringing democracy to the Middle East. Even a seasoned journalist like *New York Times* correspondent John F. Burns seemed to take these claims at face value, when he wrote in March 2004: "A month from the anniversary of [Saddam's] fall, the American project to replace him with the Middle East's first functioning democracy is in new peril." Burns implicitly accepts the premise that America's "project" in Iraq was to create "the Middle East's first functioning democracy." Similarly, Thomas Friedman, the *Times'* foreign policy columnist, wrote that "there is a 100 percent correlation of interests between America's aspirations for Iraq and the aspirations of Iraq's silent majority." Why does Friedman assume he knows what those in Washington were hoping to achieve in Iraq?

He seems to simply accept what Bush administration officials now say: that they went into Iraq to bring democracy to the Iraqi people. Just why the Bush administration would be so obsessed with helping Iraqis—spending $5 billion a month that it could otherwise put towards improving the lives of voting Americans or simply hand over to rich Americans in tax cuts—is never really explained. Friedman just assumes that it's true, and we're supposed to do the same.

The willingness of mainstream commentators to trust the motives of those in the White House is particularly odd since most commentators subscribe to modern economic theory, with its cynical view of human nature. Modern economic theory holds that the individual (or *homo economicus* as the economics textbooks call him) is motivated primarily by material self-interest. Any suggestion that humans are motivated by loftier concerns—for social justice or the well-being of others, for instance—is quickly dismissed as naïve and idealistic. Yet, oddly enough, modern economic theory isn't seen as applying to those who occupy the White House, who are presented as eschewing material concerns for higher ideals like world peace, democracy and liberating people from oppression. This, we are supposed to believe, is the "sophisticated" view.

––––––––

This shyness about suggesting the U.S. could have been motivated by a desire to gain control of Middle Eastern oil is actually something new. Before oil became a touchy subject in the run-up to the invasion of Iraq, there was a long tradition of U.S. officials acknowledging both publicly and privately that the U.S. needs oil and is willing to use military force to secure it. So, for instance, General Anthony Zinni, then commander-in-chief of CENTCOM, had no hesitation about identifying U.S. interest in Middle Eastern oil when he appeared before Congress in 1999: "With over 65 percent of the world's oil reserves located in the Gulf states . . . [the U.S. and its allies] must have free access to the region's resources." Unless the general was only kidding, or unless he meant there had to be free access to some other resource in the region—sand, perhaps—the inference was clearly that the U.S. was prepared to use military force to maintain access to the Gulf's oil.

There was certainly no reticence among U.S. policy makers back in World War II in admitting that they regarded control of oil as a vital American policy objective. "The oil resources of Saudi Arabia [are] among the greatest in the world," asserted a 1945 State department memo, and they *must remain under American control* for the dual purpose of supplementing and replacing our dwindling reserves, and of preventing this power potential from falling into unfriendly hands. [italics added]"

The determination to keep Middle Eastern oil from falling into "unfriendly hands" wasn't just about keeping other foreign powers out of the region. It was also about keeping Middle Eastern countries themselves from taking control of their own oil in a way Washington considered hostile to its interests. So, for instance, the Arab oil embargo—in which the Arab countries temporarily cut back their oil shipments to the U.S. to punish Washington for its support of Israel during the 1973 Arab-Israeli war—stirred ideas in Washington about physically taking control of Middle Eastern reserves. Such ideas were openly advocated in the U.S. media at the time. *Harper's* magazine unabashedly ran an article in its March 1975 issue entitled "Seizing Arab Oil" which advocated sending in the Marines to take over crucial oilfields. Meanwhile, President Gerald Ford and Secretary of State Henry Kissinger made clear in their public statements that, if the U.S. were seriously deprived of oil, they would not rule out military intervention.

The notion that the U.S. has the right to use force to assure its access to oil was expressed directly by Jimmy Carter, who is generally regarded as the least militaristic U.S. president in recent decades. In his 1980 State of the Union address—shortly after the fall of the U.S.-supported Shah of Iran and the Soviet invasion of Afghanistan—Carter declared that the U.S. would regard any attempt to "gain control of the Persian Gulf region" as "an assault on the vital interests of America" that would "be repelled by any means necessary, including military force." This became known as the Carter Doctrine, and it marked the beginning of a significant build-up of U.S. forces in the region and a refocusing of U.S. military strategy around the Persian Gulf. We will return to this history later. The point here is to note that Washington has been highly sensitive to its vulnerability on oil—and willing to use force if necessary to overcome that vulnerability—since the Second World War, and particularly since the Arab oil embargo in the mid-70s.

All recent administrations have focused on the U.S. vulnerability on oil, but perhaps none more than the administration of George W. Bush. The energy task force convened in 2001 by Vice President Dick Cheney is best known for producing a grab bag of tax cuts, subsidies and deregulations favouring the energy industry. But the task force also showed that the new administration was well-versed in the dilemma of dwindling world oil reserves, and particularly the nasty arithmetic of dwindling U.S. reserves. In his task force report, Cheney acknowledged that U.S. production had peaked in 1970 and that by the year 2000 "U.S. total oil output had fallen to . . . 39 percent below its peak." As a result, Cheney noted, U.S. "dependence on foreign sources of oil is at an all-time high and is expected to grow." A key message from the report, produced with a clear sense of urgency in only four months, was that "the U.S. and global economies remain vulnerable to a major disruption of oil supplies."

Cheney had been focused on the oil shortage problem even before becoming vice president. In a speech to the London

Institute of Petroleum in November 1999, while he was CEO of oil services giant Halliburton Company, Cheney zeroed in on the difficulty of finding sufficient new reserves to keep up with the growing demand for oil. "By some estimates there will be an average of two percent annual growth in global oil demand over the years ahead along with conservatively a three percent natural decline in production from existing reserves. That means by 2010 we will need on the order of an additional fifty million barrels a day. So where will the oil come from?" Cheney went on to point out that "the Middle East, with two-thirds of the world's oil and lowest cost, is still where the prize ultimately lies."

Cheney's focus on the Middle East—"where the prize ultimately lies"—certainly continued after he became vice president. "By any estimation, Middle East oil producers will remain central to the world oil security," he noted in his task force report. "The Gulf will be a primary focus of U.S. international energy policy." And, as in his speech to the London Petroleum Institute, Cheney highlighted the problem of state control of Gulf oil industries, arguing in his task force report that Middle Eastern governments be urged "to open up areas of their energy sectors to foreign investment." A year later, Cheney openly linked Washington's growing focus on Iraq to Iraq's pivotal role in the energy picture. Addressing a veterans' group in Nashville in August 2002, Cheney warned that if Saddam Hussein got control of deadly weapons, he would "seek domination of the entire Middle East" and "take control of a great portion of the world's energy supplies."

But, as the Bush administration assessed its options just before invading Iraq in the spring of 2003, the advantages of securing vast, untapped oil fields—thereby guaranteeing U.S. energy security in an era of dwindling reserves and enabling U.S. oil companies to reap untold riches—were apparently far from mind. What really mattered to those in the White House, the sophisticates assure us, was liberating the people of Iraq.

Of course, the invasion has so far failed to secure Iraq's oil for the U.S., making the sophisticates all the more convinced that oil wasn't a motive for the invasion. They're further convinced by the fact that the invasion and occupation have been very expensive—already costing Washington more than $270 billion. Why spend all that money when you could simply buy the oil from countries keen to sell it? "You just write them a cheque," argues author Gwynne Dyer.

It's a nice glib answer, but it misses the point. Given the strategic importance of oil, Washington wants its supply to be secure, not subject to the whims of hostile foreign powers. Dyer's cheque-book solution wouldn't have been much help back in 1973, when the Arab countries decided to cut back their oil exports to the U.S. for political reasons. That left an indelible impression on U.S. strategic planners who have since focused on ensuring America will never again be cut off from this most vital resource—a focus that has, if anything, become more intense in the face of dwindling worldwide oil reserves. So spending $270 billion on Iraq wasn't a calculated plan to get oil at the cheapest price. It was a calculated plan to secure Washington's control over a resource U.S. policymakers considered vital.

There was also the added benefit of spectacular rewards for a small group of private interests closely connected to the Bush administration. And for this small group—led by Dick Cheney's old firm Halliburton, but also including Bechtel, General Electric, DynCorp. and other companies that rely on war to keep the dividends flowing—the profits have already been enormous. (More on this later.) As for the major oil companies—they haven't yet been able to develop Iraq's oil; access to that treasure chest presumably lies ahead, once the military situation has been secured. In the meantime, Big Oil has been scoring unprecedented profits due to tight worldwide oil supplies. Ironically, Iraq's chaotic situation, which has kept oil production there lower than before the invasion, has contributed to the tight world oil market. So, while it is obliged to wait for a chance at the bonanza beneath the Iraqi sand, Big Oil is nicely positioned. Notes Fadel Gheit: "The oil companies are crying all the way to the bank."

So, yes, the invasion has been expensive. But who pays and who benefits? The staggering costs have been shouldered by all American taxpayers (apart, that is, from the costs borne by Iraqis: 100,000 dead, their cities and infrastructure in ruins, their country blanketed by 1,500 tonnes of deadly depleted uranium), while the staggering benefits have gone to a tiny coterie of well-connected interests. To Halliburton, Bechtel, GE, and other members of this select little group, the costs probably don't seem all that high.

———

The quest for oil has been a central feature of the fossil fuel age of the last century, and promises to be even more so in the coming century. Weaning ourselves from our acute dependence on oil would help diminish the violence, lawlessness and theft that often has accompanied this quest, if only because the black gold at the heart of the hunt would lose much of its lustre. Of course, there would be other things to fight over, other resources to secure. But oil's centrality to the modern global economy—and its finite supply and non-renewable nature—makes it a particularly desirable prize and therefore an especially potent source of conflict.

Reducing our level of dependency on oil, then, would offer a very clear benefit—not to mention the benefit of saving the ecosystem of the planet. Lined up against these sorts of momentous, civilization-enhancing advantages, it's hard to imagine why any credence should be given to concerns about whatever loss of "freedom" some people might feel having to settle for more fuel-efficient vehicles.

In other words, within reach is a far more promising scenario. Under this scenario, truly, *nobody gets hurt*—except the modern-day Luddites who have built their financial and political empires around the continuing dominance of fossil fuels. The question is: Can a tiny but enormously powerful set of private interests block the will of the entire world to move forward into the post-fossil fuel age? And the answer is—so far, at least—yes.

CHAPTER 2

ALONG COMES IRAQ

In the run-up to the U.S. presidential election of 2004, the Bush administration began to cast itself in a new role: that of innocent victim.

Things had turned out rather badly in Iraq; U.S. soldiers kept dying and, worse still for the administration, its rationale for going to war had utterly collapsed when top U.S. weapons inspector David Kay had confirmed that Iraq had no weapons of mass destruction. This could have created an embarrassing chink in the armour of the Bush administration. But as the election approached, a new theme emerged: yes, the country had gone to war under false pretenses, but no, it wasn't the president's fault. Just like the American people, the president had been the victim of faulty intelligence.

This might have just seemed like a transparent case of buck-passing. But the argument was apparently bolstered in July 2004 by the release of an exhaustive 511-page report on the failure of prewar intelligence by the Senate Intelligence Committee. The report blamed U.S. intelligence officials for grossly overstating Saddam's arsenal of deadly weapons. And the report had been endorsed unamiously by both the Democrats and Republicans on the committee. That gave it an air of neutrality and fairness: obviously the Democrats would not want to let a Republican president off the hook on such a crucial question, only months before an election.

In fact, the report sidestepped the central question: what role had officials in the Bush administration played in orchestrating the grossly exaggerated impression of Saddam's arsenal? There had been allegations that the administration, particularly the vice-president and his staff, had applied constant pressure to the intelligence agencies—the Central Intelligence Agency (CIA) and the Defence Intelligence Agency (DIA) to come up with a frightening portrait of Saddam's weaponry. If true, this changed the story considerably, making administration officials instigators rather than victims of the faulty intelligence that had been used to justify the war. But the Senate Intelligence committee had not considered this question in its report. In fact, it had specifically decided to defer this most important question to phase two of its investigation, the findings of which would only be released after the election.

In effect, this meant that there would be no attempt to hold the Bush administration to account for the war—at least not before the election, when it could make a difference. One might have thought that this would provoke a storm of opposition from the Democrats and some sharp commentary from the media. But there was little reaction.

Of course, both the Democrats and the media were somewhat complicit in the reckless rush to war, having failed to raise questions to challenge the adminisration's case. The explanation for this failure now was that everyone had been caught up in "groupthink." This was the excuse offered up by CNN's Wolf Blitzer, whose influential daily interview program would have provided an excellent vehicle to raise doubts about the White House's case for war. But Blitzer had, if anything, helped the White House make its case. Appearing on *The Daily Show with Jon Stewart* after the release of the Senate report, a regretful

Blitzer chalked up his failure to a kind of collective blindness: "We should have been more skeptical."

We should have been more skeptical? In fact, beyond the elite circle of political and media insiders that Blitzer inhabits, skepticism had been rampant about the administration's case for war. But the Democrats and the media seemed almost determined to close ranks with the Republicans in snuffing out that skepticism.

Consider, for instance, a segment on the prewar intelligence failure on NBC's *Meet the Press*, presumably one of the more hard-hitting of the network news programs. In July 2004, *Meet the Press* moderator Tim Russert interviewed the chairman of the Senate Intelligence Committee, Senator Pat Roberts, and the vice-chairman, Senator Jay Rockefeller. Roberts, a Republican, was predictably uncritical of the Bush administration. On the other hand, Rockefeller, a Democrat, expressed the view that the administration had applied considerable pressure to the intelligence community (Rockefeller had written a dissenting opinion making this point), but he seemed content to let this question go unexplored by the committee until after the eleciton.

This refusal to deal with such a crucial issue during the election campaign was so striking that even the genial Russert felt the need to put forward a few gentle probes.

"Was there any political–was there any political pressure from the White House not to do the second part of the investigation until after the election?," Russert asked.

Before the question was even completed, Roberts jumped in and assured the TV audience that Russert was way off track, that there'd been no such presssure: "None. None. None."

Roberts went on to insist that the role played by the White House in prewar intelligence would be examined, and that "even as I'm speaking, our staff is working on phase two and we will get it done."

"Before the election?" Russert asked.

"I don't know if we can get it done before the election. It is more important to get it right . . ."

Now presumably this would have been an ideal moment for Russert to express the skepticism that millions of viewers must have felt as they watched a Republican senator gently allow a Republican president off the hook on the debacle unfolding in Iraq.

But Russert didn't follow up, and Rockefeller actually came to his Republican colleague's defence. "Yeah, I absolutely agree with Pat Roberts that doing it right is more important than the 'election deadline.' . . ." Any American viewer hoping to cast an informed ballot must have been exasperated by this slow-moving investigation. The committee had been studying the issue of pre-war intelligence for more than a year, interviewing dozens of witnesses; was it really too much to assume they could give some assessment of the White House's role before election day?

The interview moved off into other territory, leaving the focus on the shoddy work done by the intelligence community, and how this had led to war. Tim Russert even suggested at one point in the show that Bush had been skeptical about the quality of the information he was getting about Iraq's arsenal but had been won over by overzealous intelligence officials. This painted the president as someone just trying to get at the truth. Any conclusion suggesting the opposite—that he and his administration were the main perpetrators of deception about Iraq's arsenal—would just have to wait until after the election, when it would become little more than an interesting historical footnote.

But, as it turned out, even assembling the facts for a historical footnote may be an uphill battle.

In addition to the investigation by the Senate Intelligence Committee, the issue of faulty prewar intelligence was also being probed by an independent commission which had been appointed by the Bush administration. This commission, headed by a Republican-leaning judge and a centrist Democrat, was also put on a schedule that assured its report would not be released until after the election. Furthermore, it was not authorized to investigate the central question of what role the Bush administration may have played in exaggerating prewar intelligence. Not surprisingly, then, when it released its report in March 2005, this "independent" commission—handpicked by the administration—found much to blame in the actions of the intelligence agencies, but had nothing to say about the role of the administration.

Of course, there was still phase two of the senate committee's investigation yet to come. Chairman Pat Roberts had, as we've seen, pledged on national TV to investigate the White House's role as part of the second phase of his committee's investigation. But, with the release of the independent commission's report, Roberts simply changed his mind. He issued a press release praising the thoroughness of the independent commission's work. "I don't think there should be any doubt that we have now heard it all regarding prewar intelligence," the Senator said in his release. "I think that it would be a monumental waste of time to replow this ground any further."

So, whatever role the Bush administration played in distorting evidence about Iraq was not going to be probed after all. Exploring whether the White House had knowingly taken the country to war on false pretenses—in other words, whether its case for war amounted to a deliberate lie—was now dismissed as "a monumental waste of time."

Later that spring, the British press broke a story about a top-secret British memo that cast new light on the mystery surrounding faulty pre-war intelligence. The memo reported details of a meeting in July 2002 attended by top British officials, including Prime Minister Tony Blair. British intelligence chief Sir Brian Dearlove, who had just returned from high-level talks in Washington, reported to the meeting that Bush wanted "to remove

Saddam, though military action, justified by the conjunction of terrorism and WMD. *But the intelligence and facts were being fixed around the policy* [italics added]." And British Foreign Secretary Jack Straw told the meeting that "the case [for war] was thin. Saddam was not threatening his neighbours, and his WMD capability was less than that of Libya, North Korea or Iran." The senior officials at the meeting, including the prime minister, then struggled with how to justify invading Iraq when such an invasion appeared to violate international law.

The "Downing Street" memo caused a considerable stir in Britain, but much less of a stir on the other side of the Atlantic. The U.S. political and media establishment apparently just didn't consider it much of a story that America's top ally thought the Bush administration was fabricating the case for an illegal war.

Certainly, the White House had resolutely tried to keep the focus on Saddam's weapons, constantly hyping the danger they posed. On October 7, 2002, Bush had even suggested that failure to take out Saddam could lead to the ultimate nightmare: "Facing clear evidence of peril, we cannot wait for the final proof—the smoking gun—that could come in the form of a mushroom cloud."

Bush had also hinted at a nuclear catastrophe in his State of the Union address in January 2003, insisting that Saddam had tried to buy enriched uranium (for nuclear weapons) from Niger. That certainly pumped up the war fever—until chief UN nuclear weapons inspector Mohammed ElBaradei scrutinized the document purporting to provide evidence of the Niger deal and declared it a forgery. A forgery! So, the most important, widely watched speech of the year by the "leader of the free world" had contained a crucial assertion justifying war, and the assertion turned out to be based on a forgery! Surely all hell was about to break loose in the world's most sophisticated democracy. But no. Quite the contrary.

There was virtually no media questioning of the forgery. Who had made the forgery? Why? How had it not been detected in this most carefully vetted of speeches? Or had it been? These fascinating questions barely registered on the media's radar screen. A ruckus did develop over whether or not it mattered that George W. Bush had presented false information to the world to justify war. Why should it matter?, the administration and its supporters asked. After all, they insisted, the false information was just a tiny part of the president's overall address, amounting to only "sixteen little words." How could sixteen little words possibly matter? Presumably, then, Bush could have used his State of the Union address to announce: "The constitution is abolished. I'm leader for life. And I plan to bomb Vermont." Hell, that's only fourteen words. He could even have added "I'm gay!" and still have uttered no more than "sixteen little words."

Over in Britain, the Blair government had also gone to great efforts to flesh out the claim that Saddam had a deadly arsenal, and there was huge fallout after the war when no weapons were found. The government reeled under allegations, made in a BBC report, that it had "sexed up" its case for war—allegations that came under intense scrutiny after the apparent suicide of government weapons inspector David Kelly. A public inquiry was held in which scores of fascinating internal government documents were unveiled, revealing just how the "sexing up" had taken place. For instance, Blair had asked his aides in September 2002 to prepare a dossier on Iraq's weapons to help win over a skeptical British public. Britain's intelligence services had in fact just completed such a dossier, but Blair's aides were dissatisfied with it. It didn't make Iraq sound very threatening at all, saying only that Iraq had the *capacity* for producing chemical and biological weapons and noting that, even if sanctions against Iraq were lifted, it would be at least *five years* before Saddam could produce nuclear weapons. (Five years! At that pace we'd have to wait forever for the tanks to roll!)

There was no way the British public could be whipped into a war frenzy with that sort of intelligence. So the officials were sent back to their drawing boards. Over the next couple of weeks they substantially overhauled the dossier about Iraq's arsenal, in a process supervised by Blair's political aides, who constantly urged that the language be made stronger. At one point Blair's press secretary apologized to intelligence officials for pushing so hard: "Sorry to bombard on this point . . ." But bombard he did. When a redrafted version of the dossier claimed that Iraq "might already have" started producing VX gas, Blair's press secretary complained about the word "might," noting it "reads very weakly." The intelligence team redrafted again, reporting back the next day: "We have been able to amend the text in most cases as you proposed." Among the changes: a claim that the Iraqi military "may" be able to deploy chemical and biological weapons within forty-five minutes was firmed up into a claim that the Iraqi military "are able to" deploy such weapons in forty-five minutes. Much better!

And so it was that the British intelligence officials' original dull but accurate portrait of Iraq's lacklustre weapons arsenal was transformed into a hair-raising account of hellish destruction that could be unleashed in less than an hour. Hence the heated headlines in the major London dailies upon the dossier's release: *45 Minutes from Attack* and *He's Got 'Em. Let's Get Him.*

The documents released at the inquiry provided an astonishing window into the way the Blair government had manipulated the facts to make the case for war. Yet the inquiry, headed by Lord Hutton, ruled that there was nothing wrong with how the government had acted, even though Hutton conceded that Blair's obvious desire for a stronger dossier might have "subconsciously influenced" intelligence officials. Incredibly, Hutton reserved his wrath for the "defective" editorial practices of the BBC, which had got some aspects of the original "sexing up" story wrong, even though the essential thrust of the story was correct. It's perhaps forgotten that the two *Washington Post* journalists who broke the Watergate scandal also made some mistakes in their reporting, which in the end did nothing to divert the public's attention from the real story—the treachery of President Richard Nixon. It's interesting to imagine what Hutton would have thought of Watergate. If we applied his standards, we'd remember Watergate as the scandal that finally exposed the sloppy editorial practices of *The Washington Post*.

————

Similarly, in the U.S., there was plenty of evidence pointing to the administration's pro-active role in generating false prewar intelligence on Iraq. Much of the erroneous information came not from the intelligence agencies but from the Office of Special Planning (OSP), which was set up by influential administration neoconservatives Paul Wolfowitz and Douglas Feith and staffed with neoconservative ideologues. The OSP keenly seized on evidence from unreliable Iraqi exiles who had little real knowledge of Saddam's weapons capability and strong motivation for exaggerating it. While the intelligence agencies certainly made serious mistakes, they had repeatedly attempted to filter out much of this unreliable information and had warned the administration of the untrustworthiness of some of its key sources.

Instead of heeding this advice, the administration actively pressured the agencies to come up with intelligence that helped make the case for war. In his dissent to the Senate intelligence committee report, Jay Rockefeller cited the evidence of one CIA veteran that "the hammering of analysts was greater than any he had seen in 32 years at the CIA." A comprehensive investigation by a team of writers, published in *Vanity Fair* magazine in May 2004, documents how vice president Cheney personally showed up at CIA headquarters on a number of occasions and clearly communicated that he wanted intelligence suggesting Iraq had a highly developed weapons capacity. He appeared to deliberately intimidate those who presented him with a less alarmist picture, the magazine reported. Even at the time, there had been media reports of unhappiness at the CIA over pressure applied by the administration. Vincent Cannistraro, a former head of counter-intelligence at the agency, was quoted in the British newspaper, the *Guardian*, saying: "Basically cooked information is working its way into high-level pronouncements and there's a lot of unhappiness about it in intelligence, especially among analysts at the CIA."

If the administration had had any desire for factual information on Iraq's weapons—as opposed to uncorroborated snippets of gossip that could be cobbled together to make a case for war—one obvious source would have been the team of UN weapons inspectors actually on the ground in Iraq. Unlike the CIA, which hadn't had any agents inside Iraq for years, the UN inspection team had been scouring Iraq since the fall of 2002. But chief UN weapons inspector Hans Blix could find no evidence of biological or chemical weapons. Blix made this point emphatically in a major presentation to the United Nations on February 14, 2003, when he contradicted key aspects of the U.S. case for war. The UN's top nuclear inspector, Mohammed ElBaradaei, also reported to the UN that day that he'd failed to find evidence of a nuclear weapons program in Iraq. As we now know, those reports by Blix and ElBaradaei were accurate.

Certainly, the inability of UN inspectors to find any weapons of mass destruction plus the doubts expressed within Washington's own intelligence agencies would have given pause to an administration that was genuinely trying to determine if

Saddam posed a threat. The solution seemed obvious: hold the bombs; give the inspections more time.

When the Bush admininstration refused to back down after Blix's powerful UN presentation, the world practically choked in disbelief. The next day, more than ten million people took part in protests around the world, including an estimated 500,000 in New York City. Without access to intelligence data, with nothing more than their own intelligence and questioning minds, millions of ordinary people in the U.S. and elsewhere figured out what top-paid journalists had been unable or unwilling to see—that there were glaring flaws in the U.S. case for war.

———

By the spring of 2005, the issue of exaggerated prewar intelligence had been officially put to rest, with barely a protest in the mainstream media. To the extent that there was any commentary on the subject, it centered around the issue of who should be blamed for the mistakes. The faulty prewar intelligence on Iraq was frequently described as one of the biggest intelligence failures in recent history, and critics correctly noted that someone should have been held accountable. And yet no one had. The president and vice-president had been re-elected (probably helped by the fact that the Senate hadn't produced a report blaming them for concocting false evidence making the case for war), Rumsfeld was still Defense Secretary, Condoleeza Rice had been promoted to Secretary of State, Paul Wolfowitz was to head the World Bank; even CIA director George Tenet, who resigned amid the harsh criticisms of the CIA under his watch, was awarded the Medal of Freedom, the nation's top honour.

The issue of accountability was undoubtedly important. But there was another important and interesting question that remained hidden below the accountability question. *Why* had the information about Iraq been falsified? If the White House was responsible for generating the false information, as appears to be the case, what was it up to? If the case for WMD was fabricated, what was being covered up? *What was the real reason for going to war in Iraq?* No commission or committee was set up to look into this crucial question.

The real motive for the invasion of Iraq was simply left unanswered. With the case for WMDs discredited—along with allegations that Saddam had had connections to al Qaeda—the White House deftly shifted emphasis to a new explanation for war: its desire to bring democracy to the Middle East. And the media seemed surprisingly willing to take this new explanation on faith, even after being badly burned by the deceptions on the earlier explanations. As the White House started ramping up talk about democracy, the media helped smooth the transition by shifting its focus onto questions like whether or not democracy was possible in the Middle East, whether the people were ready for it, whether Islam was inherently undemocratic, etc.

The word "oil" remained unmentioned and unmentionable.

———

The reluctance to acknowledge oil as a factor in the U.S. invasion is particularly odd given the readiness with which oil is identified as a factor in the behaviour of other countries towards Iraq. In the run-up to the war, for instance, North American media commentators were quick to point out—quite correctly—that France, Russia and China all had designs on Iraqi oil, and had in fact made deals with Saddam to develop the oil, once UN sanctions against Iraq were lifted. Commentators suggested that this explained the reluctance of these three countries to join the U.S.-led invasion: they already had their oil deals in place and didn't want to be obliged to hand over a piece of the action to the Americans. Fair enough. While it seems unlikely that such concerns fully explain the reluctance of these countries to join the

U.S. war effort, they were likely a factor. In any event, it was entirely appropriate for commentators to consider what role such considerations might have played.

But commentators have been strangely reluctant to acknowledge any oil-related motivation when it comes to Washington. To the extent that Washington was depicted as being at all influenced by oil, the motivation was presented as purely defensive, part of an effort to prevent aggressive actions by others aimed at getting control of the valuable resource. In his speech to the Veterans of Foreign Wars in August 2002, Cheney expressed fears of Saddam cornering much of the world's oil. "Armed with an arsenal of these weapons of terror, and *seated atop 10 percent of the world's oil reserves, Saddam Hussein could then be expected to seek domination of the entire Middle East, take control of a great portion of the world's energy supplies*, directly threaten America's friends throughout the region, and subject the United States or any other nation to nuclear blackmail [italics added]." In this grab bag of aggressive activities that Saddam is said to be plotting, oil clearly figures prominently. The suggestion is that, once he gained control of a great portion of the world's energy, Saddam would use this power to hurt the West, presumably by cutting off its access to oil.

This might seem a plausible worry at first glance. But in fact Saddam had been greatly weakened by his defeat in the first Gulf War, and was probably not plotting much of anything beyond his own survival throughout the 1990s. There was certainly no evidence that he was trying to block Western access to oil. On the contrary, Iraq had been desperately trying to sell its oil to anyone who would buy it, since oil was virtually its only source of revenue. What held it back from developing and selling a lot more oil were the extensive UN sanctions, which were maintained for more than a decade, largely at U.S. insistence. Even with this continuing U.S. hostility towards Iraq—American warplanes carried out frequent bombing missions over the country—Iraq was still very keen to sell oil to the U.S., and did so under the UN oil-for-food program. In fact, the U.S. was Iraq's largest oil customer prior to the U.S. invasion, purchasing two million barrels of Iraqi crude a day.

By presenting Washington's actions as purely defensive—as designed to ensure that Americans and their allies weren't cut off from essential oil supplies—the administration rendered the U.S. motives palatable for public consumption. Oil, if a factor at all, was presented as a benign factor. There was certainly no suggestion of any desire on Washington's part to gain control over oil in order to achieve geopolitical leverage or financial gain for American companies. Washington said it was simply trying to protect our collective interests by ensuring access to oil (even if there wasn't actually any threat to that access), just as it claimed to be protecting all of us from Saddam's weapons of mass destruction (even if those weapons didn't actually exist). This notion of a collective interest blurs the fact that U.S. actions in Iraq really serve only a very narrow set of interests.

These interests are made up primarily of two groups: first, a cabal of powerful advocates of a more assertive U.S. foreign policy, and second, a number of companies with interests in oil or other aspects of the reconstruction and redesigning of Iraq. While the oil and oil-related industries have always been interested in Iraq, the more immediate impetus behind the 2003 invasion appears to have come from the first group—the cabal favouring a more forceful U.S. foreign policy. This group consists largely of influential, radical Republicans who have long pushed for America to overcome its "Vietnam syndrome" and act more forcefully in the world (that is, even more forcefully than it always has). Some of these radicals had got their sea legs during the ultra-conservative heyday of the Reagan administration, and had later occupied important posts in the administration of the first George Bush.

With the collapse of the Soviet Union towards the end of the first Bush administration, the world political situation changed dramatically. While millions around the world were hoping that the Soviet demise would usher in a new era of peace and disarmament, the hawks inside the administration saw another opportunity: a bold new era of U.S. potency. The hawks wanted Washington to move immediately to fill the power void created by the end of Soviet power, to establish the U.S. as the lone, unrivalled superpower in the world.

Inside the Pentagon, plans were underway to seize the moment. The vision of a supremely dominant America was articulated clearly in a classified Pentagon document leaked to the *New York Times* in March 1992. The document is notable not only for its striking vision of unchecked American power, but also because of the involvement of two key figures: then defense secretary Dick Cheney and then undersecretary for policy Paul Wolfowitz.

Drafted by Wolfowitz, the document argued for Washington to develop sufficient military might to deter any nation or group of nations from contesting American supremacy or "even aspiring to a larger regional or global role." Foreshadowing later Republican hostility towards the United Nations, the document also showed a clear willingness to act unilaterally when the world body fails to co-operate with U.S. plans: "[t]he United States should be postured to act independently when collective action cannot be orchestrated." And, foreshadowing the "Coalition of the Willing," it made the case that "we should expect future coalitions to be ad hoc assemblies." Called the *Defense Planning Guidance*, the Wolfowitz document was also clear on Washington's need for access to oil. "In the Middle East and Southwest Asia, our overall objective is to remain the predominant outside power in the region and preserve U.S. and Western access to the region's oil."

The aggressive stance urged in the document caused considerable controversy, both in the U.S. Congress and in foreign capitals, and some senior administration officials denied that it was official policy. The final version of the document, released several months later, was considerably toned down. And, with the election of Bill Clinton later that year, the radical Republicans were obliged to put the implementation of their vision on hold. To their frustration, the Clinton administration didn't seem particularly interested in taking advantage of the power void created by the Soviet demise.

In 1997, a number of these radical Republicans, including Dick Cheney, Donald Rumsfeld and Paul Wolfowitz, came together to form a think-tank called the Project for a New American Century (PNAC)—by which they apparently meant a twenty-first century that would belong to a more muscular America.

The PNAC's overtly militaristic and imperialistic aims, while often cited by critics of the Bush administration, have gone largely unacknowledged in the mainstream media. To put things in some perspective, one can imagine the reaction of the Western media if a group of prominent political figures in Moscow in the 1980s had established a group called The Project for New *Soviet* Century and urged Moscow to increase its military spending in order to create an international order dominated by the Soviets. Western commentators would have been quick to point to the obvious imperialistic aims of such a group.

From early on, the PNAC had its eye on Iraq. In January 1998, the PNAC sent a letter to President Bill Clinton urging that he make the removal of Saddam Hussein an American objective. Of course, requests from think-tanks flow into the White House all the time, and this one, coming from a group of Republican hawks, created no noticeable change in U.S. foreign policy. Its significance became clear only after the administration of George W. Bush came to power in 2001, and a number of the eighteen signatories of the PNAC letter to Clinton—including Rumsfeld, Wolfowitz and Richard Perle—ended up in very senior roles in the new administration. The dream of invading Iraq no longer seemed out of reach.

Among other things, the PNAC letter to Clinton makes clear that the neo-conservatives' interest in overthrowing Saddam predates the September 11 attacks—by several years. The letter also clarifies that oil figures prominently in the PNAC's concern about protecting "our vital interests in the Gulf." These vital U.S. interests are spelled out in the letter and include "the safety of American troops in the region, of our friends and allies like Israel and the moderate Arab states, and a *significant portion of the world's supply of oil* [italics added]."

Factors other than oil are also clearly on the minds of the PNAC crowd. The security of Israel, specifically, is identified as a concern. The idea of using military force to remove Saddam had been actively pushed two years earlier by a member of the PNAC group, Richard Perle, a hard-liner from the Reagan years with close ties to Israel. In a memo drawn up in 1996 for then Israeli Prime Minister Binyamin Netanyahu, Perle advocated the removal of Saddam as part of a larger strategy for redesigning the Middle East to enhance Israel's security and allow for Israeli expansion. Perle called for the Oslo Accords, with their premise of a land-for-peace deal with the Palestinians, to be replaced by a policy of Israel permanently annexing the entire West Bank and Gaza Strip. Essential to this strategy would be the removal of regimes in the region that might offer serious resistance, which Perle identified as Iraq under Saddam as well as the regimes in Syria, Lebanon, Saudi Arabia and Iran. Perle was later appointed chairman of the Defense Policy Board, a highly influential body that advises the Pentagon.

What's striking in Perle's vision is not simply the notion that the entire region should be redesigned to accommodate Israel, but that the ultimate goal is to establish Israel's right to dramatically expand its borders to include land where millions of Palestinians currently live. This helps explain Perle's focus on Saddam Hussein as a possible threat. While Saddam posed no serious military threat to Israel (especially after Israel destroyed Iraq's one nuclear reactor in a 1981 bombing raid), Saddam was the most prominent of the few remaining outspoken critics of Israel among the Arab regimes in the region. Saddam offered open support to the Palestinian resistance, even making financial payments to the families of Palestinian suicide bombers. So, while Iraq under Saddam posed no meaningful military threat to Israel—and certainly none whatsoever to the U.S.—it did present a political obstacle to the dreams of Israeli expansion articulated by Perle.

In addition to clearing the way for Israel's expansion, getting rid of Saddam opened up the tantalizing prospect of extending U.S. influence and control over the Middle East. Saddam was an impediment to this plan insofar as he didn't defer to U.S. hegemony in the region. It was his obstructionist attitude that offended the PNAC crowd, rather than his brutal treatment of his own people. (He had been at his most brutal back in the 1980s, when he was on good terms with the U.S., even receiving friendly visits from then-U.S. envoy Donald Rumsfeld, later a PNAC member.) So, not only was Saddam potentially an obstacle to Israeli political objectives, he was an obstacle to increased U.S. dominance over the region. He was even a dangerous example. He was an Arab leader who wasn't under the thumb of the U.S., a precedent that might serve as an inspiration to Arab nationalists who had long chafed under the weak-kneed, pro-American regimes in the region—the ones the PNAC refers to as "moderate" Arab regimes.

In September 2000, just before the U.S. election that brought Bush to power, the PNAC released a major document articulating more fully its vision of American global hegemony. The U.S., it argued, must be transformed into "tomorrow's dominant force." (Of course, with the disappearance of the Soviet Union, the U.S. already *was* the uncontested dominant force, but the PNAC apparently wanted Washington's dominion over the world to be even more clearly asserted.) The PNAC noted, however, that transforming the U.S. into "tomorrow's dominant force" could take a long time, in the absence of "some catastrophic and catalyzing event—like a new Pearl Harbor." Twelve months later, the attack on the World Trade Center provided just such a catastrophic event. With many of the PNAC crowd—including Cheney, Rumsfeld and Wolfowitz—now occupying very senior positions in the White House, the possibility of bringing about regime change in Iraq seemed suddenly much closer.

Plans for overthrowing Saddam were well underway in the Bush administration months before the attacks on New York and Washington. We now know a great deal about the prominence of Iraq on the administration's agenda during its early months in office, thanks to a number of sources, particularly *Wall Street Journal* reporter Ron Suskind's book *The Price of Loyalty: George W. Bush, the White House, and the Education of Paul O'Neill.* Suskind had the close collaboration of former U.S. treasury secretary Paul O'Neill, a widely respected moderate Republican who was dismissed by the White House for failing to support Bush's tax cut agenda. O'Neill, an independently wealthy former corporate CEO, appears to bear no malice, nor does he have anything to gain from criticizing the administration. It's clear that O'Neill was shocked by what he encountered inside the White House. He described his surprise at discovering, during the first meeting of the National Security Council, only ten days after the new administration took office, that ousting Saddam had already become a top administration priority. "Getting Hussein was now the administration's focus, that much was already clear," O'Neill recalled. Indeed, the discussion of Iraq at the meeting seemed to O'Neill almost like a scripted exchange. Cheney appeared to be orchestrating things, showing "uncharacteristic excitement" as he urged those in the room to look at an aerial photo purporting to show an Iraqi factory producing deadly weapons. O'Neill wasn't convinced, and interjected: "I've seen a lot of factories around the world that look a lot like this one. What makes us suspect that this one is producing chemical or biological agents for weapons?"

O'Neill recalled that, two days later, at the next National Security Council meeting, Rumsfeld rejected a suggestion from Secretary of State Colin Powell for new targeted sanctions against Iraq, pushing instead for the actual overthrow of Saddam. "Imagine what the region would look like without Saddam and with a regime that's aligned with U.S. interests," Rumsfeld reportedly said. In O'Neill's account, Cheney again appeared to be orchestrating things.

Ron Suskind notes that by February 2001—only a month after the new Bush administration took office, and seven months before 9/11—the decision to topple Saddam seemed to have been made. "Already by February, the talk was mostly about logistics. Not the *why*, but the *how* and *how quickly*."

This focus on Iraq continued after September 11 even though there was no evidence linking Iraq to the terrorist attack. O'Neill reports that, at a National Security Council meeting on September 12, Rumsfeld argued that striking back against international terrorism would surely, at some point, require striking Baghdad. The following weekend, at a meeting of a newly created "war cabinet" at the Camp David presidential retreat, Wolfowitz urged action against Iraq. "I thought what Wolfowitz was asserting about Iraq was a reach, and I think others in the room did too. It was like changing the subject," recalled O'Neill. "I was mystified. It's like a bookbinder accidentally dropping a chapter from one book into the middle of another one. The chapter is coherent, in its way, but it doesn't seem to fit in this book."

O'Neill's account of this early focus on Iraq is reinforced in two other significant books that were published in the spring of 2004. In *Plan of Attack*, Bob Woodward reports that even before Bush's inauguration, Cheney had asked outgoing defense secretary William S. Cohen to arrange a briefing for the incoming president on top security matters, including "a discussion about Iraq and different options." And in *Against All Enemies*, Richard A. Clarke, who served as the top counter-terrorism expert in the White House before resigning in 2003, also documents a pre-9/11 focus on Iraq within the administration, and a corresponding failure to pay much attention to growing evidence of an upcoming attack by Al Qaeda.

Clarke's account adds telling detail about the speed and dexterity with which the administration switched the focus immediately after September 11 from Al Qaeda to the man in the palace in Baghdad. On the morning of September 12, Clarke arrived at the White House expecting "a round of meetings examining what the next attacks could be . . . Instead I walked into a series of discussions about Iraq." Clarke describes himself as "incredulous," and then recalls that he "realized with almost a sharp physical pain that Rumsfeld and Wolfowitz were going to try to take advantage of this national tragedy to promote their agenda about Iraq." That evening, Clarke encountered the president in the White House Situation Room. Clarke says Bush pulled him and a few others into a conference room and urged them to review everything to "see if Saddam did this. See if he's linked in any way . . ."

This focus on Iraq was in keeping with the intense interest in Iraq displayed in the PNAC's 1998 letter to Clinton—signed by, among others, Rumsfeld and Wolfowitz. And that letter was specific about the reasons for the interest in Iraq: concern over the security of Israel and American allies in the region, as well as concerns over who would control the region's oil reserves. Presumably these were still the concerns of the neo-conservative crowd now that they were in the White House.

None of this early planning for the overthrow of Saddam was publicly articulated, either before or immediately after 9/11. Rather, Afghanistan was suddenly centre stage. Within a month of the terrorist attacks, the U.S. invaded the bleak, poverty-stricken Asian nation, easily toppling the Islamic fundamentalist Taliban regime and settling into a long hide-and-seek military quest for Al Qaeda leader Osama bin Laden in the country's sprawling mountain ranges. Publicly, the focus was on Afghanistan, which seemed stuck in a fifteenth-century time warp, and also on Saudi Arabia, home of fifteen of the nineteen 9/11 hijackers. Inside the administration too, the focus was on these two baffling trouble spots, but also still very much on Iraq.

———

The Bush administration's fixation on Iraq, which was clear long before September 11, 2001, gained new momentum after that day—although not for the usually stated reasons (WMD, Al Qaeda connections, bringing "democracy" to the Middle East). Rather, the intensified interest was linked directly to oil.

In the wake of 9/11, Iraq's immense oil potential took on new geopolitical significance. One clear effect of the attacks was to highlight the seriousness of the problem that lurked inside Saudi Arabia, where so many of the 9/11 terrorists, including bin Laden himself, had grown up. This fact dramatically pushed to the fore-front a worrying scenario that had long obsessed Washington planners: the prospect of losing control of the world's biggest oil supply, located under the Saudi sand. That concern gave new energy to the drive to conquer Iraq, where the world's second-largest supply of oil is located, and the only other nation with the potential to perform the key "swing producer" role that keeps the world economy functional. (Briefly, Saudi Arabia is called a "swing producer" because its enormous reserves and highly developed oil fields allow it to quickly and dramatically increase or decrease its overall daily production of oil. Thus, it can effectively regulate the world's supply of oil, by adding some when there's a shortage and withholding some when there's a glut. As a result, Saudi Arabia can play a key role in determining the price of oil, and by extension ensure the stability of the global economy.)

The fact that so many of the 9/11 hijackers were Saudis wasn't a fluke. It reflected the problematic nature of the U.S.–Saudi relationship, which has long been a cauldron of trouble, slowly producing a lethal brew with an enormous capacity for combustion. The Saudi royal dynasty, a brutal dictatorship dressed up in royal finery, derives whatever legitimacy it enjoys among its people from its role as the protector of the Wahhabist strain of Islamic fundamentalism, the nation's official and widely practised religious doctrine. But, not content to count on the vagaries of popular support, the Saudi royal family has for decades relied on Washington to keep it in power. Under a clear understanding between Washington and Riyadh, Washington protects the Saudi royal family from external and internal threats. In exchange, Washington has been granted enormous influence over the kingdom and its management of the world's oil. While this arrangement has suited Washington and Riyadh, it has generated great dissatisfaction among ordinary Saudis, who disapprove of their government being so closely allied to a foreign power, particularly one whose actions—notably its unconditional support for Israel and indifference to the fate of the Palestinian Arabs—are widely resented in Saudi Arabia. This resentment of America is especially strong among the strict, ascetic Wahhabi clerics and their followers, who denounce American decadence and indulgence, and blame America for corrupting the Saudi royal family and leading it from the true Wahhabi path.

Only one thing needs to be added to round out this picture: endless oil money. The disaffected Wahhabis are amply provided for. The kingdom has been afraid to confront them directly and has tried to dampen down their opposition, even curry favour with them by contributing lavishly to the coffers of Wahhabi institutions. Hence, there has developed inside the desert kingdom an internal opposition that is fierce, effectively beyond the reach of Saudi authorities, full of rage against America, willing to die for Islam—and flush with cash.

The events of 9/11 highlighted just how problematic Washington's close relationship with Saudi Arabia has become. Among Saudi dissidents—the most prominent being bin Laden—the prime target is not just the White House but also the Saudi royal family, which is held in active contempt for selling its soul to the American infidel. This suggests that sooner or later the Saudi dynasty could well be overthrown from within—a scenario that has long topped Washington's list of things that could go terribly wrong on the world oil scene. "When we think about scenarios related to the world oil market, that's usually the first scenario that pops up on the screen," says Robert E. Ebel, head of the energy program at the Center for Strategic and International Studies in Washington. "What would happen in the case of the overthrow of Saudi Arabia, if the conservatives took over, and the oil was lost? . . . The world could not withstand the loss of Saudi oil . . . our economy would grind to a halt."

Washington would clearly see this as grounds for intervention. Since the late 1970s, in fact, U.S. presidents have indicated that they regard access to Middle Eastern oil as essential to U.S. national security. "All presidents refer to the American interest in access to Middle Eastern oil as vital," notes Lee H. Hamilton, a former Democratic congressman and director of the mainstream Washington-based Woodrow Wilson Center. Preserving "unfettered access to the region's oil at tolerable prices" is, according to Hamilton, a vital interest that the U.S. "is prepared to defend by whatever means necessary."

In the wake of 9/11, with Washington's control over oil and the world economy hanging by a thread that suddenly seemed more tenuous than ever before, Iraq's voluminous reserves loomed ever larger. In addition to all the other reasons Washington hawks had for invading Iraq, they now had an apparently urgent new one: the possibility of Saudi oil slipping beyond America's grasp. Iraq was the United States' best shot at a backup plan, an insurance policy against the unthinkable.

⸺

From his corner office in the heart of New York's financial district, Fadel Gheit, senior oil analyst at the prestigious Wall Street firm Oppenheimer & Co., keeps close tabs on what goes on inside the boardrooms of the big oil companies. A former chemical engineer with Mobil, the fit, distinguished-looking Gheit has been watching the oil industry closely for twenty-five years.

Gheit notes that one of the biggest problems facing oil companies today is what to do with the incredible bonanza they've accumulated in recent years. "Large and small, the companies reported record profits last year," he said. "They're sitting on the largest piles of cash in their history." (Along with Exxon's all-time profit record of $25 billion in 2004, Shell's profit soared that year to $18.5 billion and ChevronTexaco's to $13.3 billion.)

But what to do with all this cash? As mentioned in Chapter One, the financial prospects of an oil company are closely linked to finding new oil reserves. Even if wealth is simply to be maintained at its existing level, new reserves must be found to replace old reserves as they are used up. "Think of reserves as wealth," says Gheit. "[Imagine] all your reserves are in your backyard, in a swimming pool, and every day you are taking a very big pail of oil and selling it. That's how you support yourself. If you don't put new oil in the pool, in six months it will run dry. Oil companies don't want to be in this situation . . . they don't want to be less wealthy." So, to maintain their wealth, they have to find ever more oil, and the more they produce, the more they have to find to replace what's been produced. The swimming pool has to constantly be topped up. "If ExxonMobil, or any company, is producing two to three million barrels per day," notes Gheit. "In order to maintain its reserve level, the company has to *find* two to three million barrels a day, just to stay even."

This puts enormous pressure on those at the top of the big companies. When they increase their production, the marketplace rewards them by pushing up the value of their companies' stock, making their shareholders happy. But then they have to work harder at finding new supplies, to fill up the extra amount scooped out of the swimming pool.

This brings us back to the basic dilemma of dwindling oil reserves. While the companies are richer than ever, the earth's crust is poorer than ever; it is less yielding of the vital ingredient that has made them so rich. As the earth's most accessible supplies of oil have been used up, the companies have had to probe farther and deeper for new reserves—cutting through thick permafrost and going deeper and deeper underwater. Twenty years ago, the companies were drilling for oil on ocean floors at a depth of 1,000 to 2,000 feet. Now, they drill at depths below 9,000 feet. Needless to say, the costs associated with such exotic hunts are enormous, making oil from these sources extremely expensive for the companies to produce. But, despite the extra costs, they get no extra payment when they sell it. There is essentially one world price for oil—and it is the same price whether the oil costs a few dollars per barrel to produce (as in the Middle East) or $15 to $25 per barrel to produce (as in some less accessible locations). As long as the world price is high—as it's been in the last few years—even oil from marginal areas can be profitable. But, of course, this expensively produced oil is still not as profitable as the cheaply produced oil, which is getting scarcer and scarcer. "The cheap oil has already been found and developed and produced and consumed," says Gheit. "The low-hanging fruit has already been picked."

Well, not *all* the low-hanging fruit has been picked.

––––––

Nestled into the heart of the area of heaviest oil concentration in the world is Iraq, overflowing with low-hanging fruit. No permafrost, no deep water. Just giant pools of oil, right beneath the warm ground. This is fruit sagging so low, as it were, that it practically touches the ground under the weight of its ripeness.

Not only does Iraq have vast quantities of easily accessible oil, or low-hanging fruit, as Gheit calls it—but its vast store of oil is almost untouched. Although it was one of the first areas in the Middle East to be developed back in the early part of the last century, its development was never taken very far by the international oil companies—for reasons that we will explore later. Frustrated by this slow pace, the Iraqi government took matters into its own hands, nationalizing the country's oil in 1972. But before development could really proceed, the country was enmeshed in a bitter, draining war with Iran that lasted eight years and soaked up all its financial resources. Two years later, in 1990, Iraq invaded Kuwait with even more disastrous consequences. Militarily vanquished, financially crippled and unable to recover due to comprehensive sanctions, Iraq's hopes of developing its oil reserves were put on long-term hold. "Think of Iraq as virgin territory," says Gheit.

Iraq's *potential* reserves are almost certainly much bigger than is currently known. In the last twenty years, there have been enormous technological advances in computer and satellite imaging that make it possible to more accurately determine the amount of oil a country has. So, for instance, it is possible today to determine that Saudi Arabia has roughly 260 billion barrels, which amounts to roughly a quarter of the world's oil. Two decades ago, Saudi Arabia's known reserves were only about a half or even one-third of this. The amount of oil in Saudi Arabia hasn't increased, just our awareness of it. The same is true with other countries, where, in many cases, the proven reserves have increased substantially, due to more advanced exploratory techniques. But Iraq remains frozen in time, with its oil potential almost completely unexplored in two decades. Yet, even with the twenty-year-old information, we can establish that Iraq has at least 10 per cent of the world's oil. If that ends up doubling or tripling—as neighbouring Saudi Arabia's proven reserves did over the past twenty years due to improved technology—the size of Iraq's reserves could be staggering.

So it is not just that the Iraqi fruit hangs low and there is much of it to pick, but there is also so much of it yet to discover. That's the promise of the glimmering Iraqi oasis—a virtually endless supply of undiscovered, low-hanging fruit. "That's why," as Gheit puts it, "Iraq becomes the most sought-after real estate on the face of the earth."

A few quick calculations illustrate the point. Even assuming costs as high as $5 to $6 a barrel (high, that is, for such low-hanging fruit) and a world oil price of $50 a barrel (which could turn out to be a low estimate), Iraq's oil will generate a profit of well over $40 per barrel. That translates into an annual profit of $30 billion a year for every 2 million barrels pumped per day. But if Iraq ends up pumping 5 million barrels a day—which is not inconceivable, considering that Saudi Arabia pumps 8 million barrels per day—then the *profit* of Iraqi oil rises to a staggering $70 billion a year. At 7 million barrels a day, the profit rises to almost $100 billion a year, and so on. (Note that this is

roughly four times Exxon's current staggering profit of $25 bil-
lion a year.) "This is bigger than anything Exxon is involved in
currently . . . It is the superstar of the future," says Gheit. "This
is the big dance. Everybody wants to be there."

Gheit points out that there was one more factor that made
Iraq particularly attractive—its military vulnerability. Russia, for
instance, could satisfy a lot of the oil industry's need for new
supplies, since it also has huge growth potential. But Russia has a
nuclear arsenal. As Gheit notes, "We can't just go over and . . .
occupy [Russian] oil fields. It's a different ballgame." Iraq, however, was ideal—weak and, ironically, utterly lacking in weapons
of mass destruction. (If Iraq had actually had such weapons, a
U.S. invasion would have been far too risky.) Furthermore, Iraq's
central location in the Gulf area—nestled in between Saudi
Arabia and Iran, at the top of the Persian Gulf—makes it an ideal
spot for a major military presence. Even after much of the U.S.
force is withdrawn and an Iraqi government operating under
U.S. protection is established, Iraq will remain an important military staging ground from which the U.S. can exert control over
the entire region. Gheit smiles: "Think of Iraq as a military base
with a very large oil reserve underneath. . . . You can't ask for better than that."

———

There's something almost obscene about a map that was passed
around among senior Bush administration officials and the corporate executives who took part in the Cheney task force on
energy in the spring of 2001. The map (reproduced at
the front of this book) is of Iraq, and it is very detailed. But
it doesn't show the kind of detail normally shown on maps:
cities, towns, regions. Rather, its detail is all about Iraq's oil
resources. Dozens of oil fields are identified, with special desig-
nations to indicate "supergiant" oil fields (those with reserves

bigger than five billion barrels, the map's legend specifies). The southwest of the country is neatly divided into nine "Exploration Blocks." Oil pipelines are also marked, as well as refineries and tanker terminals. This map shows a naked Iraq, stripped of political trappings, with only its ample natural assets in view. It has the feel of one of those charts you sometimes see in a supermarket meat department, which identify the various parts of a slab of beef so customers can find their way to the most desirable cuts. Block 1 might be the strip loin, Blocks 2 and 3 are perhaps some juicy tenderloin, but Block 8—ahh, that could be the filet mignon.

The map may seem crass—something for oil executives to salivate over privately—but then, it was never intended for public consumption. It was one of the many documents considered by the Cheney task force, and it only became publicly available due to the determined efforts of a Washington-based public interest group, Judicial Watch, which went to court to force the Bush administration to release documents connected with the unusually secretive vice-presidential task force. Although Judicial Watch obtained the map and released it in July 2003, it generated little mainstream coverage. (Why would the fact that administration officials and their secret confidantes in private industry were studying a detailed map of Iraq's oil fields be of any interest to the public, when we all know the invasion of Iraq had nothing to do with oil?)

Another fascinating document obtained and released by Judicial Watch was a two-page chart titled "Foreign Suitors for Iraqi Oilfields." It identifies sixty-three oil companies from thirty countries and specifies exactly which Iraqi oil fields each company is interested in and the status of the company's negotiations for those oil fields with Saddam's regime. So, for instance, we learn that Canada's Ranger Oil Ltd. was interested in "Block 6, other" and had "Signed MOU [memorandum of understanding] with Baghdad." Another Canadian firm, Chauvco Resources Ltd.

of Calgary, was interested in the Ayn Zalah oil field, and there had been "[a]dvanced talks by late 1996. Service contract for advanced oil recovery (gas injection project) in this aging field." Among the other companies identified are Royal Dutch/Shell and Russia's Lukoil, which had apparently made substantial headway in lining up deals with Saddam for Iraq's abundant north Rumaila oil field. Also listed is France's Total Elf Aquitaine, which had its eye on the fabulous Majnoon oil field, with its reputed 25 billion barrels. Baghdad had "agreed in principle" to the French company's plans to develop this succulent slab of Iraq. There goes the filet mignon into the mouths of the French.

These documents clearly suggest that the Cheney task force was very interested in Iraq's oil fields and who would get to develop them after sanctions against Iraq were lifted. (Additional documents released to Judicial Watch show the task force also had detailed oil maps of Saudi Arabia and the United Arab Emirates.) But what makes the Iraqi oil map significant is the fact that, as we now know from many accounts, the Bush administration was actively focused on invading Iraq right from its first days in office (and long before 9/11). In other words, at the same time that the White House was intensely focused at the highest levels on a possible invasion of Iraq, it was also keenly studying Iraq's oil fields and assessing how far along competitors were in their negotiations with Saddam for a piece of the Iraqi oil action. Dick Cheney himself appears to have been masterminding both the push to invade Iraq and the task force deliberations.

The central role of Dick Cheney in these two initiatives—both launched almost immediately after the new administration took office—is noteworthy, particularly given his enormous power within the administration. The fact that Cheney was strongly focused on the invasion of Iraq and, at the same time, on energy policy is certainly suggestive of a possible connection between the invasion and a desire for oil. Fadel Gheit, the Wall

Street oil analyst, describes the two as "very connected." Yet this connection is the very thing that is always denied. This might explain why the White House seems so keen to present the invasion of Iraq as Bush's idea, not Cheney's. In fact, this is what is most striking about Bob Woodward's account of the run-up to the war in Iraq. Woodward was given extensive access to the top people in the White House and an astonishing three and a half hours with Bush himself. His account can in many ways be regarded as the official White House version of what happened. While it provides many intriguing details, it presents a somewhat unconvincing portrait—based largely on Bush's own account—of Bush as the driving force behind the invasion. Interestingly, Bush wasn't even a signatory to the PNAC documents in the 1990s about the importance of getting rid of Saddam, and was reportedly not much focused on foreign policy at all before deciding to run for president. Yet Woodward seems to accept the notion that it was Bush who was pushing to topple Saddam. Woodward even goes to the trouble of trying to refute the fairly widely held perception that Cheney is the real force behind the scenes. Naturally, the White House wants to make Bush look strong and presidential. But it also might be keen to divert attention—on the question of who was the driving force behind the invasion of Iraq—away from Cheney, who has a particularly strong focus on energy, and specifically oil. Again the aim seems to be to disconnect the invasion from oil.

Back to the meat chart. What it points to is the *commercial* nature of the interest in Iraq. Whatever concerns there may or may not have been about Iraq's weapons, about connections to Al Qaeda or any larger geopolitical considerations, the documents obtained by Judicial Watch suggest that there was also a strong interest in Iraq's oil wealth, and in the danger of it falling into the hands of eager foreign oil companies—rather than into the rightful hands of eager U.S. oil companies. Indeed, as the document about "foreign suitors" shows, non-U.S. oil companies already had their feet well in the door and were nicely positioned for major involvement in Iraq once UN sanctions were lifted. By contrast, the major American companies, after years of U.S.–Iraq hostilities, were largely out of the picture, and would have been the big losers if sanctions had simply been lifted—a development that seemed increasingly likely, given the international pressure against the sanctions and the sheer logistical difficulty of maintaining them. "The U.S. Majors stand to lose if Saddam makes a deal with the UN [on lifting sanctions]," noted a report by Germany's Deutsche Bank in October 2002.

James A. Paul of the New York-based Global Policy Forum agrees, and argues that losing out on Iraq would have amounted to a serious blow for the future prospects of the U.S. oil companies. "All those other companies would get the prize deals," says Paul. "That was the whole future of the oil industry." Iraq's oil fields are the big remaining bonanza to be harvested and, had they fallen into the hands of other companies, Paul insists that, over time, the impact would have been devastating for the U.S. majors: "[They] would have been ruined in the international oil industry. They wouldn't have had [the] reserves." Paul argues that the presence of China's national oil company among the foreign suitors was particularly provocative to Washington, which sees China as the big future threat to U.S. hegemony.

Presumably, the disadvantaged position of U.S. oil companies in Saddam's Iraq was discussed at meetings of Cheney's task force—and with representatives of those very U.S. oil companies. We can't say for sure exactly who might have been in on such discussions because, incredibly, the task force refused to reveal any details about what went on in its meetings, or even the names of those who attended or with whom it consulted. Legal action has tried to pry out some information. Among the litigants has been the General Accounting Office (GAO), the investigative arm of

Congress, which sued in federal court for access to the task force's records. A GAO report in August 2003 concluded that the task force had relied on advice from oil industry officials—it mentioned oil giant Chevron—as well as officials from the coal, gas and nuclear industries. Press reports have also pointed to Big Oil's involvement with the task force. *Time* magazine reported that the task force met with nearly fifty energy companies or associations, and specifically mentioned that Conoco president Archie Dunham, whom *Time* decribes as "an old Cheney pal," visited the vice president on March 21, 2001, to present Conoco's views.

The extreme secrecy that surrounded the consultations lends credence to the notion that the oil company executives were discussing, among other things, how to prevent foreign competitors from beating them to Iraq's oil (which would explain why they were examining the meat chart and the list of foreign suitors). In some public statements, oil companies had advocated simply lifting the sanctions against Iraq (and the other "rogue states" like Libya and Iran), so it is conceivable that they were privately pressing Cheney just to lift sanctions rather than to bring about actual regime change in Iraq. (Recall, however, that the Deutsche Bank predicted that the end of sanctions would leave U.S. oil companies in a disadvantaged position, clearly something the executives wouldn't have been keen on.) Since we know that Cheney was, in the same time period, actively pushing within the administration for the overthrow of Saddam, it certainly seems likely that toppling the Iraqi regime was an idea that Cheney raised in his meetings with the oil executives. The executives must have been at least comfortable with the idea; if not, their objections would have carried great weight. Indeed, it is hard to imagine Cheney and Bush proceeding with plans to invade Iraq over the objections of Big Oil, a constituency that had long been very close to both men and had provided them with extensive financial backing.

The administration's close ties to the oil industry are legendary and well documented. As the non-partisan Washington-based Center for Responsive Politics has noted, George W. Bush, who was involved in running several oil company ventures in the 1970s, received more money from the oil and gas industry in the 1999–2000 period than any other U.S. federal candidate had received *over the past decade.* ExxonMobil alone contributed $1.3 million. (In addition to the industry's ties to Bush, Cheney served as CEO of Halliburton until his vice-presidential selection, and Secretary of State Condoleeza Rice served on the board of Chevron throughout the 1990s.) The extensive nature of the oil industry's ties to and financial backing of the Bush White House makes it hard to imagine that the administration's policies did not substantively reflect the desires of the industry. As we've seen, the administration's plans to invade Iraq were up and running as soon as it took office, so one might reasonably assume that they had been formulated in the pre-election period—the very period when Bush was receiving all that money from the industry. Once in office, Bush immediately began implementing two key policies that would impact enormously on the oil industry: withdrawing from Kyoto and taking control of Iraq. The industry's massive financial support certainly suggests that it probably had significant input into both of these policies.

It seems likely, then, that Big Oil was actively involved from the very early stages in the Bush team's plans for Iraq, and that great efforts were made to keep this involvement secret—precisely because it would taint the invasion with the smell of oil, which was above all what the administration wanted to avoid. Exactly what role Big Oil may have played in the planning is unclear. Fadel Gheit suggests simply that the companies "can direct the antenna of the administration about what should be considered."

One intriguing piece of evidence pointing to Big Oil's involvement appeared in an article in *The New Yorker* magazine in

February 2004. Staff writer Jane Mayer reported *
National Security Council (NSC) document from February 200*
directed NSC staff to co-operate fully with Cheney's task force,
which had just been established. This might seem odd on the face
of things, since the task force was focused on energy policy while
the NSC is concerned with military and defense issues. But the
NSC document noted that the task force would be considering
the "melding" of two policy areas: "the review of operational
policies towards rogue states" and "actions regarding the capture
of new and existing oil and gas fields." This implies that the
Cheney task force was looking beyond the normal domain of
domestic energy policy, that it was also considering geopolitical
questions about how to "capture" oil and gas reserves in "rogue"
states, including, presumably, Iraq.

If there's a smoking gun, that might be it. Mark Medish, an
NSC official from the Clinton years, says the document raises
questions about what the Cheney task force was up to. "Why
were they rolling out maps of Iraqi oil fields in February 2001?"
he asks, noting that the public believed the secretive Cheney
task force was dealing exclusively with domestic issues. "But if
this little group was discussing geostrategic plans for oil," said
Medish, "it puts the issues of war in the context of the captains
of the oil industry sitting down with Cheney and laying grand,
global plans." In other words, it appears that Big Oil, through
the Cheney task force, was involved in discussions about a U.S.
intervention in Iraq. Indeed, it appears that Cheney and his former colleagues from the oil industry were developing a
comprehensive energy policy that included the possibility of
taking control of Iraq. This certainly helps explain the otherwise baffling level of secrecy surrounding the task force, which
was supposedly only considering domestic energy issues that
Big Oil would normally be consulted on anyway. Since, according to the NSC document, the task force was to focus on bigger

geopolitical questions such as getting control of oil in Iraq—
something Big Oil would want to be distanced from—there was
a compelling need for secrecy.

There is a similar veil of secrecy over Big Oil's involvement
later on, in the fall of 2002, once the plans for invading Iraq
were well developed and out in the public domain. For instance,
in October 2002, Ahmed Chalabi, the Iraqi exile leader whom
the Bush administration originally hoped to install as leader in a
post-Saddam Iraq, held secret meetings with three multinational
oil companies to discuss future oil contracts. When reports of
the meetings began leaking out, Chalabi's exile group confirmed
them, explaining to journalists: "The oil people are naturally
nervous. We've had discussions with them, but they're not in the
habit of going around talking about them." Why were the oil
people "naturally nervous," and why were they holding secret
meetings with Chalabi anyway? He had long been pushing for
the overthrow of Saddam, and had developed close ties to
Washington hawks. It seems that he was also trying to enlist the
support of the big oil companies, and they appear to have had at
least sufficient interest to meet with him. Certainly his plans for
post-Saddam Iraq would have pleased them. As Chalabi told *The
Washington Post*, he favoured the establishment of a U.S.-led consortium to develop Iraq's oil, and "American companies will have
a big shot at Iraqi oil." (Not surprisingly, this kind of talk alarmed
executives in non-U.S. oil companies, who feared being squeezed
out of future contracts by a Chalabi-run government. Lord
Browne, chief executive of British-based BP, said at a briefing on
the company's financial results that he had pressed Washington
not to allow Iraq to be carved up exclusively for U.S. oil companies in the aftermath of war. "We have let it be known," he said,
"that the thing we would like to make sure, if Iraq changes regime,
is that there should be a level playing field for the selection of oil
companies to go in there if they're needed to do the work there.")

Meanwhile, another secret meeting was apparently held between Big Oil and top U.S. war planners. *The Wall Street Journal* reported that Cheney's staff met in October 2002 with executives from ExxonMobil, ChevronTexaco, ConocoPhillips and Halliburton, and discussed plans to secure and rehabilitate Iraq's oil fields. The *Journal*, which has strong connections in the business world, reported that it had learned of this meeting from "industry officials." It also reported that both the Bush administration and the oil companies denied the meeting had taken place. It's possible, of course, that *The Wall Street Journal*'s industry sources were mistaken, that the meeting never took place. Another possibility is that it did indeed take place but was denied by both the administration and the companies because of the extreme sensitivity about any suggestion that invading Iraq was about oil.

Poking around in this story, one is struck by the extent to which the footprint of Big Oil has been removed. In all the reams of space the media has devoted to the invasion and occupation of Iraq, there's been little said about oil, and even less about Big Oil. Well out of sight is any notion of the kinds of financial rewards— in the range of $70 to $100 billion a year, as mentioned by Fadel Gheit—that could someday be reaped by private oil companies. Such riches are simply not part of the story, as the media instead present the public with endless tales about the difficulties faced by the U.S. as it helps Iraq make the transition to democracy.

One reason that regime change in Iraq was seen as offering significant benefits for Big Oil was that it promised to open up a treasure chest that had long been sealed—private ownership of Middle Eastern oil. Prior to the 1970s, the oil industries of the Middle East were privately owned and operated by a small group of major international oil companies, which paid a royalty to the host government for oil they extracted. But starting in the early

1970s, most Middle Eastern countries (and Venezula) nationalized their oil industries. As a result, state-owned oil companies today control the vast majority of the world's oil. The major international oil companies control a mere 4 percent.

The majors have clearly adjusted to the new system, and have prospered under it, since they've continued to dominate the worldwide refining and marketing of oil. And they've also continued to do much of the actual drilling and pumping in the Middle East, although they now do it under contracts with the host governments. They've been reduced to the role of developers rather than owners, and there's little doubt that they'd prefer to be owners in the oil world's Garden of Eden. "[O]ne of the goals of the oil companies and the Western powers is to weaken and/or privatize the world's state oil companies," observes New York–based economist Michael Tanzer, who advises Third World governments on energy issues. (Dick Cheney also alluded to the oil industry's interest in privatizing state-owned oil companies in his speech to the London Petroleum Institute in 1999, but noted that "progress continues to be slow.")

Certainly, the possibility of Iraq's oil being reopened to private-sector ownership—with the promise of truly stupefying profits—attracted considerable interest in the run-up to the U.S. invasion. Indeed, in February 2003, while Secretary of State Colin Powell was doggedly trying to convince the UN Security Council that Saddam Hussein posed an imminent threat to world safety, other parts of the U.S. government were busy developing secret plans to privatize Iraq's oil industry (among other Iraqi assets). A confidential 100-page contracting document, drawn up by the U.S. Agency for International Development with help from officials in the U.S. Treasury Department, lays out a wide-ranging plan to replace Iraq's largely state-run economy with a free market system. The document, obtained by *The Wall Street Journal* after it was circulated among private financial consultants, makes clear that oil is to be a key part of this "Mass Privatization Program." It calls for "private sector involvement in strategic sectors, including privatization, asset sales, concessions, leases and management contracts, *especially in the oil and supporting industries* [italics added]."

The U.S. officials who drafted the document clearly had no qualms about proposing this and other far-reaching changes for Iraq, along lines that would make the country a safe place for foreign investment. They call, for instance, for the establishment in Baghdad of a "world-class" stock exchange, with U.S.-trained stockbrokers, so that shares in the newly privatized companies can be traded. They also call for a redesigned Iraqi tax system "consistent with current international practice," including a consumption tax (which would, incidentally, allow the tax burden to be placed most heavily on Iraqi consumers rather than on foreign companies operating in the country).

At the same time, the Pentagon was also working on plans for opening up Iraq's oil sector in a post-Saddam era. In the fall of 2002, months before the invasion, the Pentagon had retained Philip Carroll, a former CEO of Shell Oil Co. in Texas, to draft a strategy for developing Iraq's oil. Carroll's plans apparently became the basis of a proposed scheme, which became public shortly after the war, to redesign Iraq's oil industry along the lines of a U.S. corporation, with a chairman and chief executive and a fifteen-member board of international advisers. Carroll was chosen by Washington to serve as chairman, and was sent to Baghdad as the administration's top oil adviser. But Carroll quickly grasped the sensitivity of the oil issue in Iraq, and understood that any overt plans to proceed with privatization would be highly provocative. "Nobody in their right mind would have thought of doing that," he later said.

The resolve to avoid overt privatization grew in the months following the invasion, as the anti-U.S. insurgency quickly gathered steam. Carroll left his post in September 2003, but his replacement, Rob McKee, another prominent former Texas oil executive, also understood the sensitivity of the issue and requested Washington rethink its plans. By December 2003, a new State department plan had been drafted, with heavy input from the oil industry. Overseen by Amy Jaffe, a well-connected energy consultant who later went on to work for former Secretary of State James Baker, the plan laid out seven possible models for oil production. They all involved Iraq's oil remaining under the nominal control of a state-owned oil company. But they varied widely in terms of where actual control would lie. In the model clearly favoured by the Jaffe plan—with its heavy input from U.S. oil companies—operation and control of the oilfields were to be largely handed over to foreign oil companies.

So, even if full-scale privatization was understood to be too provocative, more subtle forms of privatizing Iraq's oil continued to be on the agenda of U.S. oil companies. Robert Ebel, head of the energy program at the Center for Strategic and International Studies and a former vice-president of Dallas-based oil exploration company Enserch Corporation, was careful to stress in an interview that it would be up to Iraqis to determine how to develop their own oilfields. But he made it clear that the preference among U.S. oil executives was for Iraq to abandon its rigid nationalization of the past. Ebel, who remains close to the industry, said that the major oil companies are prepared to invest the $35 to $40 billion that will be needed to develop Iraq's reserves in the coming years. "We're looking for places to invest around the world. You know, along comes Iraq, and I think a lot of oil companies would be disappointed if Iraq were to say, 'We're going to do it ourselves.'"

Along comes Iraq? How fortuitous. U.S. oil companies just happened to have tens of billions of dollars that they wanted to invest in undeveloped oil reserves when Iraq presented itself, ready for invasion.

Along comes Iraq, indeed.

———

With just days to go before Christmas 2004, the finance minister of Iraq, Abdel Mahdi, held a press conference at the National Press Club in Washington that sparked considerable interest among oil executives and energy consultants. Mahdi, a prominent Shi'ite who lived in exile in France during much of Saddam's reign, had cultivated close ties to Washington in recent years. Already one of the leading figures in the U.S.-backed Iraqi government, Mahdi was widely expected to remain important after the elections, scheduled for the following month. (After those elections, Mahdi became one of two vice-presidents of Iraq.)

Oil industry insiders who took time out from Christmas events to hear Mahdi's talk in Washington were not disappointed. Flanked by U.S. undersecretary of State Alan Larson, Mahdi explained that Iraq was looking at privatizing its oil industry. This was delightful news; the very thing the U.S. industry had long pushed for was now being presented to them as an Iraqi idea! The burly Iraqi finance minister, known as a keen advocate of free markets, also said that Iraq would reconsider the oil deals that Saddam had signed with French and Russian oil companies. "So I think this is very promising to the American investors and to American enterprises, certainly to oil companies," Mahdi told the room full of oil company insiders.

Conspicuously absent was the mainstream U.S. media. Only a reporter from Inter Press Service, an international news service focused on development issues, was there to record the event for the public, and his account didn't get picked up by any major U.S. news outlets. Perhaps the American media just hadn't been aware of the event. After all, it was held at an obscure location: the National Press Club.

———

Of course, the occupation of Iraq has turned out to be much more chaotic and unruly than expected. The strength of the anti-U.S.

resistance has meant, among other things, that Iraq's vast oil potential remains undeveloped, and will likely remain so for a while. In the meantime, however, there is much rebuilding to do over there, and there are many major U.S. firms keen to take part.

Whatever the eventual outcome of the invasion and occupation of Iraq, one thing was clear from the outset: this was going to be a very costly venture, and huge fortunes were going to be made.

CHAPTER 3

THE MAN TO SEE

In late February 2003, as the U.S. moved inexorably towards war, top officials of the Bush administration kept their public statements focused on the evils of Saddam Hussein and the threat that his weapons posed to the world. Taking their cue from the White House, the media too kept their focus trained on Saddam and his weaponry. Meanwhile, inside the Pentagon, even with war preparations at a fever pitch, there was a focus on more practical matters.

One very practical matter was deciding which major corporation would get the contract for putting out oil fires during the invasion and getting Iraq's oil flowing after the fighting stopped. It was a very desirable contract worth potentially billions of dollars, and there were a number of interested and qualified companies in the running. So selecting the right contractor was a major item on the agenda of a high-level meeting convened at the Pentagon on February 26th, 2003. In attendance were several dozen officials from the Defense and State departments and other key sectors of the administration. Among the officials was Bunnetine (Bunny) Greenhouse, the highest-ranking civilian in the U.S. Army Corps of Engineers, the body charged with awarding Pentagon contracts.

A 60-year-old black woman, Greenhouse had managed to make her way up through the ranks of the male-dominated Corps through sheer hard work and determination. Her thorough, no-nonsense approach had attracted the attention of the Corps' first black chief engineer, General Joe Ballard, who, in the late 1990s, had been keen to break up the old-boy network that traditionally played such a large role in the awarding of contracts. Ballard's efforts were aided by new laws enacted to ensure that contracts were awarded fairly and that some of the Pentagon's largesse was directed towards smaller firms, including minority-owned businesses. He had placed Greenhouse in charge of seeing that that was done, giving her responsibility for signing off on all Pentagon contracts worth more than $10 million. There was a lot of resistance to her role inside the Pentagon, and after Ballard left in 2000, her job got even more difficult.

It got tougher still after the Bush administration came to power in 2001. Whatever reformist zeal had existed inside the Pentagon and the Corps soon disappeared, allowing the old boy network to make a comeback. With nothing more than the law to back her up, Greenhouse had been finding it harder and harder to hold the line. This meeting promised to be particularly challenging.

Greenhouse was well aware that KBR, a subsidiary of Halliburton Company, had the inside track—as preposterous as that seemed. One would have thought that the close connection between Halliburton and vice president Dick Cheney would have made it politically awkward to award such a large government contract to a Halliburton subsidiary. If nothing else, there would be the appearance of a possible conflict of interest. As was well known, Cheney had headed Halliburton before becoming vice president; his five-year stint as CEO had earned him $35 million in compensation, making him a very wealthy man. Cheney insisted that he severed his ties with the company when he left to join the Bush administration. Yet the ties didn't appear to be totally severed. In addition to whatever loyalty he might have felt to a company that had rewarded him so handsomely in the past,

Cheney had continued to receive roughly $150,000 a year in deferred compensation from Halliburton, and to hold millions of dollars in the company's stock options. Furthermore, in 2003— three years after he'd officially severed his ties with the company—Halliburton invoked an insurance policy indemnifying Cheney against potentially ruinous legal bills in connection with a multi-billion lawsuit over asbestos claims brought against Halliburton's past and current executives.

And there was an additional reason that it seemed wrong for this contract to go to KBR (formerly known as Kellogg, Brown and Root). The previous fall, the Pentagon had paid KBR $1.9 million to draw up a contingency plan for putting out oil fires during an invasion of Iraq. It was accepted practice that the company drawing up the contingency plan be excluded from getting the job. After all, whoever drew up the contingency plan would know the exact job specifications, and would therefore have an unfair advantage. That was deemed unacceptable; indeed there was a strict protocol against it. It was exactly the sort of thing that Greenhouse was supposed to prevent from happening.

As the meeting unfolded, Greenhouse was disheartened to see that there was, however, a virtual consensus in the room that KBR should get the contract, known as Restore Iraqi Oil (RIO). But she simply wasn't prepared for something more unsettling—the casual entrance of several KBR representatives at the meeting. They simply took their seats at the table and the discussion continued. Nobody but Greenhouse even seemed to think it inappropriate to have KBR representatives taking part in discussions about whether their company should be selected for a contract worth billions of dollars. As the meeting continued, it became clear that the RIO contract was an exclusive deal for up to five years. This was even worse than Greenhouse had expected. It was one thing to rush through an emergency arrangement to put out oil fires in the next month or so; it was quite another to give a company a monopoly on the work of restoring Iraq's oil wells for the next half-decade. But everyone at the meeting seemed to be in accord—no company met the terms spelled out in KBR's contingency plan nearly so well as KBR itself.

Greenhouse whispered in the ear of Army Corps Lieutenant General Carl A. Strock, who was chairing the meeting, that he should make the KBR people leave. Strock did so. But the next day, Greenhouse was asked to approve the final contract awarding RIO to KBR. Against her advice, the contract was a five-year deal, free of competition, that would simply reimburse KBR for everything it spent, with a percentage profit added on top. This "cost-plus" arrangement was notorious for allowing contractors to inflate costs—as KBR had done with a similar contract involving troop support in the Balkans. The RIO contract was potentially worth $7 billion—and the war hadn't even begun.

Still, with the pressure of war looming, Greenhouse withdrew her objections and signed off on RIO. The following month, there was a low-key announcement that the Army Corps of Engineers had awarded KBR a contract to put out oil fires in Iraq. With the war about to begin, the contract announcement went largely unnoticed.

That the administration's most influential proponent of the war had a close connection to a company reaping billions from that war was effectively treated as irrelevant. In the absence of a memo directly linking Cheney to Halliburton's multi-billion dollar war profits, the media seemed willing to give the whole issue a pass. Even when such a memo surfaced, the media all but ignored it.

———

The RIO contract had, however, attracted the attention of Judicial Watch (the Washington-based public interest group which had uncovered the oil maps of Iraq used by the Cheney task force, mentioned in the last chapter). Judicial Watch made repeated requests to the Pentagon for more information about the RIO contract but its requests were rebuffed. After obtaining a court order under Freedom of Information legislation, the group was eventually given reams of irrelevant documents, along with heavily-blacked out memos that were largely indecipherable. Among just such a pile, Judicial Watch found one intriguing email; it appeared that someone had neglected to block out some rather telling information.

The email was dated March 5, 2003—three days before the announcement that the RIO contract had been awarded to KBR. The email had been sent by someone in the Army Corps, although the names of both the sender and recipient were blacked out. It noted that high-ranking Pentagon official Douglas Feith had been given "authority to execute RIO" by deputy Defense secretary Paul Wolfowitz, and that Feith had approved the contract, "contingent on informing WH [White House] tomorrow." Interestingly, the email then went on to say: "We anticipate no issues *since action has been co-ordinated with VP's [vice president's] office*. [italics added]"

A plain reading of this email would lead one to conclude that the vice president's office had been consulted in some way on the multi-billion-dollar RIO contract. What else could be meant by "action has been co-ordinated with VP's office"?

Christopher Farrell, director of investigations at Judicial Watch, was stunned. The email seemed to provide evidence of the very thing that the administration always vehemently denied— that Cheney had in some capacity been involved with the decision to channel billions of dollars in government funds to his former employer. And there was only five months to go before the presidential election. Surely the media would go wild with this.

But, as with the oil maps of Iraq, which had sparked surprisingly little media interest, so it was with the email about Cheney and the RIO contract.

Judicial Watch initially gave the story to *Time* magazine. In its May 30, 2004 issue, *Time* reported on the email and its contents in a small sidebar that ran with a larger story on government contracting. The sidebar story included a denial from a Cheney spokesman who insisted that Cheney had "played no role whatsoever in government-contract decisions involving Halliburton." The story also quoted a Pentagon source explaining that the message in the email simply referred to the fact that "in anticipation of controversy over the award of a sole-source contract to Halliburton, we wanted to give the vice president's staff a heads-up."

But that's not what the email says. The email doesn't say that the vice president's office has been "notified" of the action. It says that the action has been "co-ordinated" with the vice president's office. Perhaps the author of the email misspoke. But, given the many billions of dollars involved and the close connection between Halliburton and one of the very top officials in the administration, one would have thought a little further probing was in order. Surely the media would pick up on the story, at least ask a few follow-up questions—like why the email says that action was "co-ordinated" with the vice president's office, if it wasn't? But there was little media follow-up. The story never went much beyond the little sidebar in *Time*.

In late October 2004—only a couple of weeks before the election—*Time* returned to the RIO contract. This time, the story was about Bunny Greenhouse's attempt to stop the Army Corps from awarding it to KBR. Greenhouse's account had surfaced because she had been demoted, and a letter from her lawyer protesting her situation had been leaked to the magazine. *Time's* report detailed her account of KBR officials attending the Pentagon meeting about the contract—striking information suggesting an unusually close relationship between KBR and the administration. But, oddly, the *Time* report mostly focused on Greenhouse's role as a whistleblower and the problems she was having as a result—rather than on what her whistle-blowing was bringing to light. The article also made no reference to the Pentagon's March 5 email about the RIO contract—which the magazine had itself reported only a few months earlier and which seemed to lend credence to Greenhouse's implicit suggestion of high-level favouritism towards KBR. Instead, the *Time* article implied that the issue here was Greenhouse. It concluded by noting that "her career and reputation are on the line."

No doubt that was true. By taking on the Pentagon, she was risking both career and reputation. But was that really the key story here? Somehow, the career and reputation of the second-highest person in the U.S government—whom voters would have a chance to pass judgment on in a few weeks—had faded into the background.

———

Before going any farther with the tale of Cheney, Halliburton and war profiteering in the rebuilding of Iraq, it is worth quickly noting something that is almost never considered: Iraqis are quite capable and keen to rebuild Iraq themselves.

This is perhaps so obvious that it's barely worth mentioning. Of course, Iraqis would want to rebuild their own country, even apart from the fact that unemployment (estimated to be about 70 per cent) is one of the biggest problems they face in their war-ravaged land. Yet this obvious possibility seems to have been completely ignored in Washington's rush to award contracts to Halliburton and other big U.S. firms. Not only would it undoubtedly be much cheaper to have Iraqis and Iraqi firms do the bulk of the reconstruction work, it might even contribute to that most elusive of qualities in Iraq these days: goodwill towards America.

It is surely one of the great ironies that the U.S. adamantly maintained before the war that Iraqi scientists and engineers had the expertise to build the most advanced weaponry, yet these same people are somehow deemed inadequate to the simpler task of rebuilding their country's basic infrastructure. It's worth noting that Iraq has a highly educated workforce—probably the most educated in the Middle East—with more than enough expertise to get its electricity and water systems working and rebuild its roads, bridges and buildings. Iraq's professionals are in fact well schooled in rebuilding infrastructure; they've been doing it for years, even when UN sanctions left them without the most basic parts, tools and equipment.

The Iraqi rebuilding skills are most evident in the nation's oil industry, according to Issam Al-Chalabi, who served as Iraq's oil minister in the late 1980s. (He should not be confused with Ahmed Chalabi, the Iraqi exile whom Washington had hoped would lead post-Saddam Iraq.) Issam Al-Chalabi is typical of the kind of technocrats who ran Iraq's oil industry under Saddam. Educated as an engineer at University College in London, he speaks fluent English and has long tilted towards the west. As Iraq's oil minister, he favoured the use of western technology rather than the less reliable Soviet technology encouraged by Saddam. Over lunch in Amman, Jordan, Issam Al-Chalabi explains to me that he was not involved in the political side of the Iraqi government, but was purely a technocrat interested in developing the wealth of the nation. He was in fact fired from his post in August 1990 (when Saddam abruptly decided he'd rather have his son-in law running the oil ministry), and now runs a prosperous international oil consulting business from Jordan.

Al-Chalabi is incensed at the notion that the likes of Halliburton are necessary to put out oil fires and rebuild Iraq's oil wells. "During the Iran-Iraq War, all the oil installations were targeted, rocketed and bombed, but we were able to rebuild and recon-struct," says Al-Chalabi, who served in several senior positions in the country's oil bureaucracy before rising to minister during that war, which raged from 1980 to 1988. The Iraqi professionals proved equally adept at rebuilding after the heavy bombing in the first Gulf War in 1991. "All the refineries were hit, all the power plants. But within six weeks, people were getting electricity and oil product," he says. "The Americans were surprised to find, after the [1991] Gulf War, that refineries built fifty or fifty-five years ago were still running efficiently—despite the sanctions and no spare parts."

Despite the billions spent on contracts for U.S. firms to rebuild Iraq since the U.S. invasion, basic services, such as electricity, remain unreliable. The U.S. forces have even been ineffective at protecting oil installations and pipelines around the country, which are constantly subjected to sabotage by insurgent forces, making it difficult to even restore oil production to prewar levels.

Al-Chalabi notes that sabotage had always been a problem, with various factions or separatist movements trying to disrupt the flow of oil or gain control of it for political purposes. To deal with this constant threat, the Iraqi oil ministry had maintained its own police force. Several thousand strong, the "oil police" had their own weapons and vehicles, and operated completely separately from Iraq's regular police forces. Their sole function was the protection of the nation's oil installations and the thousands of miles of pipelines. In addition, the oil ministry sometimes made arrangements for tribes to carry out local security in particularly dangerous areas. With these sorts of systems in place, Al-Chalabi says, Iraq had largely succeeded in protecting its oil—a task that has proved tricky for the Americans.

While the U.S. has relied on American companies to rebuild Iraq's oil operations and to provide security, there was no need to turn to Americans for these tasks, according to Al-Chalabi. "The running of the industry could have been taken over by [Iraqi] professionals. They have been doing this for decades. They are among the best in the world . . . I do not accept the [U.S.-led] coalition taking over these installations."

Al-Chalabi's assessment of the competence of Iraqi oil professionals is backed up by outside observers. A twenty-page analysis of Iraq's oil operations prepared by Germany's Deutsche Bank in October 2002 concluded that the state-owned Iraqi National Oil Company (INOC) would be well equipped to handle Iraq's oil once UN sanctions were lifted. According to the report, the INOC "has a strong track record of reserves and technology management in its past, although [it] has been seriously hampered by the politics of stop-start production, and a lack of money and technology since 1990. While the leadership of INOC could change post sanctions, the basic organizational framework, the production databases and technical staff could well form the basis of a new management company."

Above all, Al-Chalabi seems saddened and frustrated by the squandering of Iraq's enormous economic potential. He had witnessed, from the inside, just how close Iraq had come to realizing the dream of becoming a thriving modern economy, only to see it all slip away.

Oil has always been close to the centre of Iraq's hopes for national development, and in the 1970s the economic prospects looked extremely promising. With the nationalization of the country's oil industry in 1972 and the dramatic rise in international oil prices, Iraq's 25 million people, while living under brutal dicatatorship, were enjoying a high standard of living, with free medical care, free education (including university) and no income taxes. And that promised to be only the beginning; incredibly, *less than 1 percent* of Iraq's massive oil resources had been developed. As president of Iraq's State Company for Oil Projects in the late seventies, Al-Chalabi was poised to oversee the development of the country's immense oil reserves and the transformation of Iraq into the economic powerhouse of the Middle East. By 1979, the plans were all in place; the engineering and construction contracts had all been signed. Many of the contracts were with foreign firms, but all the decisions had been made in Iraq, and the Iraqi oil ministry was going to be running things. "We knew what we had. We were going to do it ourselves," he says wistfully.

Saddam's invasion of Iran in 1980 brought those plans to a halt, as all Iraq's energy and focus shifted for the next eight years to fighting the war against its neighbour. At the end of the carnage, in 1988, an ultimately victorious but financially crippled Iraq still had spectacular oil potential. Once again optimism briefly surfaced, only to be abruptly cut off with Saddam's invasion of Kuwait two years later. This time Iraq suffered a quick and decisive defeat by U.S.-led forces, followed by more than a decade of UN economic sanctions, which left much of the country sunk in poverty.

Now, with Saddam gone, Iraq's oil potential might get back on track. But the dream of Iraq's vast oil reserves developed by Iraqis—a dream that seemed well within reach back in 1979—seems once again to have slipped out of view. Al-Chalabi believes the Americans will continue to control Iraq. "They're there to stay for many, many years to come . . . I don't believe the Americans will ever leave Iraq."

———

Iraqis rebuilding Iraq and running its oil sector may be a popular dream among Iraqis, and may have seemed like a sensible idea to observers in international banks, but it was never envisioned by those in Washington who planned the American occupation. Of course, there were to be plenty of low-level jobs for Iraqis, but the management and ownership of the firms involved in the reconstruction of Iraq—and the decisions about what a reconstructed Iraq would look like—were to be almost exclusively in American hands. This had been clear from the original privatization plans (mentioned in the last chapter) drawn up by officials in the U.S. Treasury Department and the U.S. Agency for International

Development. To get the privatization scheme underway, an initial contract had been awarded—before U.S. troops had even landed in Iraq—to a Virginia-based consulting firm, BearingPoint Inc. (formerly KPMG Consulting). Subsequent contracts were to be opened up to "a limited pool of competitors," according to a report in *The Wall Street Journal.* In that limited pool, the *Journal* noted, would likely be such prominent U.S. firms as Deloitte Touche Tohmatsu and International Business Machines (IBM). The *Journal* summed up the situation: "The execution of the [privatization] plan . . . will fall to private American contractors working alongside a smaller team of U.S. officials."

As it turned out, the privatization plans envisioned prior to the war had to be altered when the resistance proved fiercer than expected. Specifically, as noted earlier, it was decided early on to remove oil from the assets to be privatized. But there was no retreat from Washington's general plan to impose privatization and foreign ownership on the country—before the Iraqi people had a chance to vote on these sweeping changes. In its 2004 budget plan, the U.S.-appointed administration in Iraq signalled its intention to open up the country's heavily nationalized economy to private investment, saying that the model of state-owned enterprises has proved "a failure across the world." Whether or not the Iraqi people agreed with this didn't matter; the issue had already been decided for them. So, while Bush repeatedly spoke about bringing democracy to Iraq, some of the most significant decisions an electorate ever gets to make—about the nature of its economy—had already been decided by officials in Washington.

Washington also took control of the huge amounts of money involved in reconstructing Iraq, not all of which came from U.S. taxpayers. In May 2003, only a month after the fall of Baghdad, Washington convinced the UN to accept its plan to move billions of dollars in Iraqi oil revenues from before and after the war into the newly created Development Fund for Iraq. A UN advisory board was set up to monitor the fund, but ultimate control was granted to the U.S.-led Coalition Provisional Authority. The actual money was to be held in the Federal Reserve Bank of New York, and a senior U.S. Treasury Department official, George Wolfe, was appointed to advise the fund, which was charged with dispersing money for, among other things, oil infrastructure and electricity programs.

Most of the money for Iraq's reconstruction has been controlled directly by the Pentagon, with hundreds of millions of dollars paid to big American firms like Bechtel, General Electric and DynCorp. But no company has received more government business from the U.S. invasion and occupation of Iraq than Halliburton and its subsidiary KBR, which have been awarded multiple contracts to provide logistical support for the U.S. Army as well as to repair Iraq's oil fields. Altogether, these contracts are worth a staggering $18 billion. Halliburton CEO David Lesar has tried to downplay the financial benefit of the contracts, insisting that, after expenses and taxes, the actual profit margins have been small. But the contracts have accounted for nearly all of KBR's recent growth. Indeed, without the contracts, KBR and Halliburton would be in rough financial shape, in large part due to Cheney's decision back in 1998 to merge Halliburton with a Texas energy company that was about to be inundated with lawsuits over asbestos claims. The Iraq contracts have managed to offset these huge liabilities.

Indeed, for a company otherwise facing a financial crunch, Halliburton has rebounded nicely in the wake of the Iraq war. But there have been plenty of charges that it has taken advantage of the government's largesse. Government auditors have issued at least nine reports criticizing the company's work in Iraq. One Pentagon audit found that KBR was unable to account for more than $1.8 billion. Several criminal investigations have been launched into possible overcharging and kickbacks on KBR contracts. KBR denies any wrongdoing.

Six former KBR employees have come forward and told Congress and the media about what appear to be examples of extreme overcharging by the company. Marie de Young, a former KBR logistics expert, described how the company put up its staff in luxurious five-star Kuwaiti hotels and built gyms and recreation centres for them, facilities that have nothing to do with the troops that the company is supposedly there to serve. De Young also maintained that KBR and a subcontractor were overcharging for laundry services—by about $1 million a month. Another former KBR employee, James Warren, who drove trucks for the company in Iraq, described how brand-new $85,000 trucks were simply abandoned or "torched" if they got a flat tire. Warren brought this and other wasteful spending to the attention of senior management, and was fired a few weeks later.

Despite these apparently egregious abuses of government money, Halliburton has received surprisingly little flack. Congressman Henry A. Waxman has managed to get Congress to devote some attention to the contracts, but the Republican majorities in both houses have shown little interest in launching further investigations. And the government has largely ignored the concerns raised by its own auditors.

All this brings us back to the question of favouritism, and the possibility that the company has received special treatment because of its relationship to Dick Cheney. Let's go back to the meeting, mentioned in the last chapter, that the *Wall Street Journal* reported took place in October 2002. The meeting, which the White House denies ever happened, was said to involve members of Cheney's staff and executives from a number of oil companies, including ExxonMobil, ChevronTexaco, ConocoPhillips—*and Halliburton*. Well, if the presence of Exxon executives in the vice president's suite wasn't enough to provoke official denials that the meeting took place, the presence of executives from Halliburton certainly would have been. Was this perhaps part of the way matters involving Iraq were "co-ordinated with VP's office," as suggested in the March 2003 Pentagon email?

The vice president denies he has played any role in the awarding of these contracts, and KBR backs up his denial. But is it likely that Halliburton's enormous good fortune—in winning contracts in Iraq and being largely immune from punishment when caught apparently abusing those contracts—has nothing to do with its ties to perhaps the most powerful man in the Bush administration?

In her article in the *New Yorker*, mentioned in the last chapter, Jane Meyer reports on a number of Republican-connected business ventures that received contracts in Iraq. She quotes an unidentified businessman, whose firm was awarded several such contracts, as saying: "Anything that has to do with Iraq policy, Cheney's the man to see."

———

In the interconnected world of the oil industry and the U.S. government, no relationship is cozier than that among Cheney, Halliburton and successive Republican administrations. After doing graduate work in political science, Cheney started out in Washington as a political aide, working in the Nixon White House under Donald Rumsfeld, who then headed the Office of Economic Opportunity. The office had played an important role in Lyndon Johnson's War on Poverty—a war that didn't much interest Nixon and certainly didn't interest Rumsfeld or his protégé Cheney. In what was to become a lifelong collaboration and friendship, Rumsfeld and Cheney together set about undermining the effectiveness of the anti-poverty office they were running, largely by getting rid of most of its staff and handing over its functions to private contractors. Cheney and Rumsfeld were thus pioneers in the game of privatization, before it became the right's favourite tactic for enriching the private sector and undermining any government program aimed at addressing inequality.

Cheney ran successfully for Congress, serving as representative from Wyoming for eleven years, during which he consistently promoted the interests of energy companies (in addition to fiercely opposing abortion under any circumstances). Chosen to be secretary of defense in the administration of George H.W. Bush, Cheney presided over the first U.S. war against Saddam, in 1991. During his years running the Defense Department, Cheney launched a bold plan to privatize a significant part of the U.S. military—a plan that, as things later turned out, was to make Cheney a very rich man. The idea was to hand over to the private sector the job of providing logistical support for military operations abroad. Traditionally, these logistical tasks—preparing meals, doing laundry, cleaning toilets, pretty much everything but actually fighting—had been carried out by reserve soldiers. Cheney proposed turning all these functions over to a single private company.

That led the Pentagon to Halliburton, a Houston-based conglomerate that had made a reputation for itself in oil-well construction and servicing, and which, through its engineering subsidiary Brown & Root, had built a good part of the army infrastructure during the Vietnam War. The Pentagon paid Halliburton $3.9 million to come up with a plan for how a privatized system might work. Halliburton was then paid another $5 million to do a follow-up study. In August 1992, with Cheney still serving as defense secretary, the U.S. Army Corps of Engineers chose Halliburton to take on the massive job that the company had just been paid almost $9 million to define.

Three months later, Bill Clinton was elected president, and Cheney was soon back in the private sector. For the next couple of years he spent most of his time making speeches and raising money across the country, as he seriously contemplated a run at the presidency. Among his major contributors were top people from Halliburton and Bechtel. Cheney eventually decided not to run. Then, in 1995, he was hired as CEO of Halliburton—despite the fact that he had no previous experience in the oil industry and no experience running a company, let alone a Fortune 500 company with worldwide operations. Clearly, his Pentagon contacts were seen as more than compensating for these apparent deficiencies. During the five years Cheney served as CEO, Halliburton won $2.3 billion in Pentagon contracts, all stemming from the army privatization scheme that Cheney himself had put in place as defense secretary. (This amount was almost double the value of government contracts Halliburton had received during the previous five years.) In addition, the company received $1.5 billion in federal loans and insurance subsidies, compared with only $100 million in the previous five years. Cheney had worked the Pentagon privatization scheme from both ends, and Halliburton rewarded him handsomely.

But the connections between Cheney, Halliburton and the Republican administrations were to get tighter still. While he was still CEO of Halliburton, Cheney headed George W. Bush's vice-presidential search committee in the spring of 2000. Under Cheney's stewardship, the committee searched high and low for a suitable running mate for Bush, only to conclude that they were staring him right in the face—or rather in the mirror. Apparently with no embarrassment, Cheney decided he was the best man for the job.

What can be made of all this? For one thing, Cheney is not someone overly sensitive to the appearance of conflicts of interest. He wasn't concerned about the propriety of paying Halliburton to draw up a plan for a project it then won, or about later becoming CEO of that company. Nor did he flinch from selecting himself for the job he was asked to find the best person to fill. Cheney surely takes the gold medal for political chutzpah.

But his penchant for self-dealing should raise questions about more than just his character. For instance, what role did he play in the U.S. decision to invade Iraq, and in whose interests was he operating? Cheney was one of the most vigorous proponents of overthrowing Saddam. It's reasonable, then, to ask: in pushing for the war, was he operating, to some extent, on behalf of a private company that stood to make billions of dollars if such a war proceeded? Would Cheney even perceive this as a conflict of interest?

It's interesting to recall here that the Bush administration's planning about Iraq began at the very beginning of its term in office. After former treasury secretary Paul O'Neill made this revelation, the administration tried to downplay its significance, insisting that this was simply a continuation of the anti-Saddam policy which had been in place during the Clinton years. But in fact, there was a key change. Clinton's policy had essentially been to contain Saddam through sanctions; Bush's was to overthrow the Iraqi leader. So this was a new policy, yet it was already in play as soon as the Bush team arrived, suggesting that its genesis must have been before the Bush inauguration. As mentioned in the last chapter, this indicates that the idea of toppling Saddam likely began to take shape while Bush was receiving record-breaking financial contributions from the energy industry for his presidential campaign. It also suggests that the idea of overthrowing Saddam may have begun to take shape while Cheney, working with the Bush team in search of (ultimately) himself, was still CEO of Halliburton. So it's possible that, in his characteristic self-serving fashion, Cheney was helping develop plans for a war that would hugely benefit a company he was not just closely connected to *but was actually running.*

Although a small number of corporate interests are benefiting enormously, the U.S. invasion of Iraq has clearly damaged the broader public interest, causing thousands of deaths and unleashing a new wave of anti-American hatred across the Middle East. This apparent willingness on the part of Washington to champion private oil interests at the expense of the broader public interest is not, however, a new phenomenon.

Washington's relationship with Big Oil is, of course, complex, and there are many factors that determine an administration's policies relating to oil. On a basic issue like oil prices, for instance, there are competing interests to balance. Oil companies generally want high oil prices, and they are powerful players and major financial contributors to political campaigns. But consumers want low oil prices, and consumers are a hugely important constituency, including not just the voting public but also major oil-consuming industries (which also make financial contributions to political parties). Indeed, the whole economy is affected negatively by high oil prices. So, no matter how generous the oil companies are in their political contributions, there are limits to what an administration can do to keep them happy.

So Big Oil doesn't always get its way in Washington. Not always, but often—despite the fact that when it does the public interest is usually hurt. As the late John M. Blair, an oil economist and historian, wrote: "[T]he historical role of the federal government has been not to restrain the industry but to make more effective its exploitation of the public interest." Certainly, the farther away from the home front, the easier it is for Big Oil to influence Washington's decisions, as it consistently did in shaping key aspects of U.S. foreign policy throughout the last century—a story we will explore in more detail later. A couple of examples here will suffice to illustrate the point, and help place the Iraq saga in a larger historical context.

―――――

Back in the late 1940s, Saudi Arabia's King Ibn Saud had little notion of the kind of wealth his kingdom happened to be sitting on. And for good reason. He had granted the concession to develop his country's oil to Aramco, a consortium of U.S. companies (Exxon, Mobil, Texaco and SoCal), which was enjoying an astounding profit rate on its investments of about 50 percent. Aramco was keen to keep the king and his kingdom happy, and so it built highways, hospitals, schools, even set up a welfare fund in the country. One thing Aramco didn't give the Saudis, however, was any information about their own oil. Aramco did all the exploring, developing, producing and marketing of Saudi oil, and so it knew a great deal about the country's sole asset, and how it fit into the international oil scene. The Saudis, however, were kept in the dark.

That darkness began to lift in 1948, with the light coming from faraway South America. In oil-rich Venezuela, the government had just scored a major victory, successfully pushing the oil companies to pay the country a 50 percent royalty on each barrel of oil. This was a major breakthrough for an oil-producing nation, and having secured it, the Venezuelans turned their attention to making sure that the oil-producing nations of the Middle East also drove better bargains with the industry, to prevent the companies from simply shifting more production over there. A Venezuelan delegation travelled to the Middle East with financial documentation—translated into Arabic—about the oil companies and how enormously rich they were getting by exploiting Middle Eastern oil. There was much information for the Saudis to feast on. First, there was the stunning news that the Venezuelans were to receive a 50 percent royalty; the Saudis were getting only 12 per cent. The Venezuelan documents further showed that Aramco was earning three times more annual revenue from Saudi oil than was the Saudi government. Aramco was also paying more in taxes to the U.S. government than it was paying in royalties to the Saudis.

If all this weren't provocative enough, another development around the same time got the Saudis thinking seriously about the possibility of negotiating better terms with Aramco. There was a huge tract of desert land between Saudi Arabia and Kuwait, officially designated a neutral zone, that was controlled jointly by the two countries. The area, previously occupied mostly by nomadic Bedouin tribes, began to take on great significance in the late forties, when preliminary oil exploration indicated the area had great promise. Up until this point, oil exploration and development in the Middle East had largely been carried out by a handful of major multinational oil companies based in the U.S., Britain and a few other European countries. But the vastness of the riches to be earned in the Middle East oil game had led to an influx of "independent" companies—that is, companies not affiliated with the "majors," which operated as a tightly run cartel. (The cartel was sometimes known as the Seven Sisters, since it involved the seven big international oil companies, based in the U.S. and Europe—more on this later.) As the Saudis began to consider the development of the neutral zone, they found themselves approached by some of these independents, and they quickly discovered that the independents were prepared to be a whole lot more generous than Aramco.

The Saudis eventually awarded the concession for the neutral zone to an independent firm that offered (in addition to building schools, housing projects and a mosque) a 55 percent royalty. It soon became clear that the independent, a small American company called Pacific Western, could pay this dramatically higher royalty and still make a very handsome profit. In fact, within eight years Pacific Western's owner, an eccentric recluse named John Paul Getty, would be the richest man in the world. (Even so, legend has it that Getty had a pay phone installed in his seventy-two-room mansion in case any of his guests were tempted to make long-distance calls.)

By 1950, it was abundantly clear to the Saudis that Aramco could be a lot more generous and still take home staggering profits. After all, oil was selling for $1.75 a barrel, but the Saudi government was receiving only 21¢ a barrel in royalties, leaving Aramco with $1.54 a barrel. (Even after all expenses and U.S. taxes were paid, the company still retained 91¢.) The Saudis began to insist on a bigger cut, and they even won support from Aramco officials located in Saudi Arabia. The executives who ran Aramco's parent companies back in New York weren't so quick to agree, however. They wanted to keep the Saudis happy, but they preferred to do so with lavish gifts (perhaps a new desert mall or a gold-trimmed mosque?), not a larger share of the profits. Clearly, they weren't keen on renegotiating the terms of what was surely the most lucrative commercial deal in world history. Also, they feared that giving the Saudis more would only encourage other Middle Eastern governments to reconsider *their* deals with the companies. The parent companies ultimately concluded there wasn't a lot that needed fixing in their existing deal with Saudi Arabia.

In Washington, experts within the State Department also pondered the situation. It was clear that the Saudis wouldn't be easily placated; the evidence that they were entitled to a bigger share was simply too strong. The problem then became how to give the Saudis more without taking away from Aramco. This seemingly unsolvable puzzle turned out to be easily solved. In fact, the desired result was achieved with only a few small changes in the Saudi–Aramco deal. Here's how the new arrangement worked: Instead of paying the Saudis a 12 percent royalty per barrel, Aramco would pay the Saudis a 50 percent royalty per barrel—a hefty increase. But the extra amount wouldn't actually be considered part of the royalty, but rather a tax paid to the Saudi government. The difference was crucial. If it was a tax paid to a foreign government, then it could be deducted, dollar for dollar, from taxes the company owed to the U.S. government. It was an ingenious little move. Whether considered a tax or a royalty, it was still the same amount of extra money from the point of view of the Saudis. From the point of view of Aramco, the change meant no change in its bottom line; Aramco would pay more to the Saudis but would then save that same amount in U.S. taxes. In fact it was much better for Aramco, because now its prime customer would be very happy. And keeping the Saudis happy was crucial to the long-term success of Aramco and its parent companies in New York.

So both the Saudis and Aramco benefited enormously from the new arrangement: the Saudis got a lot of extra money and Aramco came out the same financially while managing to please the Saudis and thereby ensure the future of the world's most lucrative deal. The only party that did badly in the deal was the taxpaying American public, which now had a large chunk of money missing from its national treasury—almost $50 million the first year. Within a few years that annual loss had risen to more than $150 million. Over the next couple of decades, as similar arrangements were made for oil companies operating in other Middle Eastern countries, the losses to the American treasury became staggering, amounting to billions of dollars. Indeed, the system created an enormous inducement for oil companies to invest abroad to take advantage of this hugely generous tax break. By 1973, the big five U.S. oil companies were earning two-thirds of their profits abroad, and not paying any U.S. tax at all on these earnings. The result was that Americans had to pay higher taxes (or accept cuts in government services or greater government indebtedness). But since the only aggrieved party—the American people—didn't know anything about the arrangement until later, this didn't turn out to be much of a problem.

Can this arrangement be said to have been in the broader public interest? Some have argued that Americans (and indeed other Westerners) benefited from the arrangement because it assured them access to cheap oil. On the surface this sounds reasonable. Americans certainly wanted cheap oil. But was their access to it ever at risk? Whether the concession was in the hands of Aramco or in the hands of another company or group of companies, Americans would still have had access to low-cost oil. (At this point the Saudis had no control over the price of oil, which was set by the major companies.) Indeed, if cheaper oil had been the main goal, then the best course of action would have been to introduce competition into the oil market by letting the independents into Saudi Arabia. This would have flooded the market with oil, bringing down the price—and in the process it would have weakened the monopoly power of the cartel operated by the major oil companies. Most people, particularly consumers, would have regarded this as a good thing. But of course, it was not what the majors wanted.

Interestingly, the possibility of allowing the Saudi concession to fall into the hands of independents was apparently never even considered by Washington. The potential benefit of the increased competition must not have been seen as important or relevant. This might be explained partly by the fact that the key State Department official who worked out the new arrangement, George McGhee, was a former vice-president of Mobil—a background that likely gave him a particular perspective on the type of solution desired. In explaining his thinking later to a Senate committee, McGhee insisted that the deal had been necessary because "[t]he ownership of this oil concession was a valuable asset for our country." Of course, most of the independents were also American. Besides, even if keeping the asset in Aramco's hands was considered best for America, why didn't the State Department simply urge Aramco to pay a more generous royalty to the Saudis?

Some have argued that this diversion of U.S. tax dollars to

Saudi Arabia was a form of foreign aid. At one of the hearings, a senator asked McGhee if the scheme hadn't been "an ingenious way of transferring many millions by executive decision out of the public treasury and into the hands of a foreign government treasury without ever needing any appropriation or authorization from the Congress of the United States?" Anthony Sampson, author of *The Seven Sisters: The Great Oil Companies and the World They Shaped*, adopts this line of thinking, arguing that the deal allowed Washington to channel foreign aid to the Saudis without having to seek Congressional approval—something that he says would have been difficult to obtain at a time when Israel was struggling for survival. But it would be wrong to construe the Saudi–Aramco deal as foreign aid. The State Department officials figured out a way to channel more money to the Saudis only because the officials wanted to ensure that Aramco retained the concession. They knew that if more money wasn't offered, the Saudis would turn to an independent or group of independents who would offer them the terms they wanted. Essentially, then, the new arrangement was simply a business deal to ensure that the highly lucrative concession stayed in Aramco's hands.

Rather than being a form of foreign aid or a way to guarantee Americans access to cheap oil, the arrangement amounted to nothing more than an intervention by Washington to prevent the Saudis from extracting more advantageous terms from the four biggest American oil companies. Without the State Department's intervention, the companies would have had to give a larger share of the profits to the Saudis or risk losing the concession—something they weren't foolish enough to do. The bottom line is that the companies—Exxon, Mobil, Texaco and SoCal—would have had to pony up the extra dollars and would not have suffered too badly for doing so. Thanks to the State Department (represented by a former Mobil senior executive), the companies were spared having to do that. They got to keep their financial boondoggle—and it fell to U.S. taxpayers to come up with the extra money to keep the Saudis happy. Far from being "an ingenious way of transferring many millions" from the public treasury to a foreign government, it was an ingenious way to transfer many millions from the public treasury to the coffers of the four American oil companies—a gift from the American people to the giant American oil companies, arranged by a government that, theoretically at least, represented the people.

This ingenious movement of money bears some similarity to the way Washington is moving money around today in Iraq. Once again, obfuscation abounds. As the Bush administration pushed through its $87-billion package for rebuilding Iraq in the fall of 2003, opposition to the bill focused on the notion that this was far too generous a foreign aid package. (In fact, about $67 billion went to support U.S. troops, leaving only about $20 billion for Iraqi reconstruction.) "It's always tough to sell foreign aid at home . . ." said Danielle Pletka, a senior analyst at the arch-conservative American Enterprise Institute, as she described the difficulty the Bush administration had winning support for its reconstruction package.

Selling Americans on foreign aid may well be difficult; certainly one can understand why Americans would want to see some of that money invested in their own country, whose infrastructure has been sadly neglected after decades of tax cuts. But it's hard to see how what's going on in Iraq could be considered foreign aid. First of all, the U.S. is largely reconstructing what it itself destroyed through more than a decade of sanctions sandwiched between two wars. And the lion's share of the money is, in any event, going to enrich a handful of U.S. companies.

The U.S. insistence on relying on big American companies to

rebuild Iraq makes no sense if the goal is simply reconstruction—something the Iraqis could handle at a fraction of the cost. No doubt the U.S. has more sophisticated technology and expertise for the rebuilding task, but what if it were left up to Iraqis—not just the grunt jobs, but the management (and ownership, if it's to be privately done) as well? Would they do a second-rate job? It's their country. Why not let them get their electricity running, their security operational, their oil into production, in their own way, as good or bad as that might be (especially since, if nothing else, there would probably be less sabotage and resistance)?

The movement of money in Iraq is not unlike the movement of money in the Aramco deal. Back then "foreign aid," apparently directed to the Saudis, actually ended up in the coffers of Aramco and its parent companies. Today, vast sums of taxpayer dollars, ostensibly directed to a foreign country (Iraq), are actually ending up in the coffers of powerful corporations (Halliburton, Bechtel, GE, DynCorp.) with close ties to the Bush administration.

A curious form of foreign aid indeed.

As with Aramco in the 1950s, Washington has often acted in the interests of Big Oil even when doing so directly conflicted with the broader public interest. In such cases, claims that "national security" was at stake were key to getting the public onside. Certainly, the invocation of "national security" has worked wonders in creating the illusion that there really are no competing interests, just a shared interest in protecting the nation. This line of argument has worked even when the claim of "national security" has been far-fetched or downright silly.

So, for instance, "national security" was invoked by Washington in the 1950s to justify a move it made to protect Big Oil from facing competition at the hands of independent American oil companies in the U.S. domestic market. The independents had long been a thorn in the side of Big Oil. Back in the 1930s, the presence of large numbers of independents producing copious amounts of oil, particularly in Texas, had threatened to undercut the price set by the major oil companies. The majors had managed to limit this competition by convincing state governments, notably Texas, to impose restrictions on the amount of oil produced within their boundaries. These production restrictions had kept up the price of oil, just as the majors had hoped. But by the late 1950s, the majors faced a new problem: independents were importing large amounts of cheap oil from the Middle East, thereby bringing down the price in the U.S. market. The majors disliked this new competition intensely, so they urged Washington to introduce restrictions on foreign oil imports and, in March 1959, a system of import quotas was imposed.

A few things are noteworthy here. Once again we find a key official with a close connection to the major oil companies: Robert B. Anderson, the U.S. Treasury secretary at the time and a leading figure of the cabinet committee that recommended the adoption of the import quota system. Just prior to accepting the Treasury post, Anderson had made a $1-million deal (involving the sale of oil leases) with prominent oil interests, including those connected to the Rockefellers. The arrangement—brought to light by *The Washington Post*'s Bernard Nossiter in July 1970—included a payment of $270,000 to Anderson during his time in office and an additional $450,000 that depended on the fate of oil prices. Among the oil interests involved in the deal were Standard Oil of Indiana and an entity called International Basic Economy Corporation, a company closely controlled by the Rockefeller family.

It's also important to note that the impact of the foreign import quota was to drive up oil prices in the U.S. The obvious benefit this offered the major oil companies was obscured by the fantasy that the import quota had something to do with "national security." During the extensive congressional hearings and debates over the import quota system, "national security" was a constant theme, indeed really the only theme. But how did keeping out foreign oil imports protect the national security? There were essentially two arguments. First, by keeping foreign oil out, it was said that Washington would avoid the danger of having foreign oil supplies cut off in the future. (True: by cutting them off now, there would be no danger of them being cut off in the future.) Second, it was argued that excessive foreign oil imports would so weaken the U.S. oil industry—a robust, thriving industry if ever there was one—that the whole industry might collapse, causing terrible damage to the entire U.S. economy!

These far-fetched arguments held the day, with little dissension. And so it was that for the next fourteen years, while the import quotas were in effect, Americans paid much more for oil than consumers in other parts of the world, in the name of "national security." The absurdity of this was nicely captured years later at a legislative hearing during testimony by an Exxon vice-president. The chief economist for the Senate Subcommittee on Antitrust and Monopoly, John M. Blair, commented to the Exxon vice-president: "Some of us have difficulty in trying to understand how the national security of the United States is strengthened by a discriminatory pricing practice under which prices paid by American buyers tend to rise while prices paid by European and Japanese competitors are steadily declining. To some of us it would seem that this would *weaken* rather than strengthen the national security of the United States."

Of course, in retrospect some might argue that higher oil prices were not a bad thing, since they encouraged less oil use. But discouraging oil use was not the intention at the time. Nor did the policy have that effect; the higher prices did little to discourage American consumption of oil, which was still relatively cheap. What the foreign import quotas did do, however, was

ensure that Americans consumed American oil, rather than foreign oil, at a much more rapid pace than would otherwise have
been the case. When the energy crisis struck in 1973, leaving
Americans scrambling for oil, Washington quickly cancelled the
import quotas to allow as much foreign oil as possible to flow
into the country. But, after fourteen years of relying almost
exclusively on their own oil, Americans discovered that their
once ample reserves were sadly depleted.

It's interesting to note the role played by the major oil companies in this depletion. Despite evidence in the mid-1950s that
U.S. oil reserves would soon be in decline, the major oil companies repeatedly denied this, insisting there would be plenty of
American oil well into the future. "Had it not been for their reassuring implications, public policy might well have been the
reverse of that actually followed from 1959 to 1973," Blair noted.
"Instead of accelerating the depletion of our limited reserves by
a forced reliance on domestic production, public policy could
have been directed towards the rational objective of conserving in
peacetime our limited supply by using foreign imports."

Washington's willingness to allow U.S. oil reserves to be
depleted was a policy fiasco of enormous proportions. Without
such depletion, the U.S. would be far more self-sufficient in oil,
and far less dependent on foreign oil. Washington's insistence on
compensating for its depleted reserves—by ensuring it has unfettered access to foreign oil—continues today to have huge
consequences for the entire world.

———

Oil figures prominently in any plan for achieving dominance in
the world economy, hence the intense interest in Iraq by those
planning a "new American century." But Iraq is only one piece of
the international oil picture. Another important piece is OPEC, the
much-vilified Organization of Petroleum Exporting Countries,

the alliance of oil producing nations that managed to push up the price of oil dramatically in the mid-1970s. In reality, OPEC performs many of the functions in today's oil markets that used to be carried out, prior to the 1970s, by the much more tightly run cartel operated by the majors. But there's a big difference—at least in the eyes of Washington—between a cartel run by Big Oil, which is seen as benign, and one run by a bunch of developing countries, which is seen as a threat to the West.

Ever since OPEC's emergence as an important player on the world stage in the mid-1970s, Washington has tried to undermine its effectiveness and weaken its unity. This strategy has been helped along by OPEC's own internal difficulties—the challenge of maintaining unity among a group of countries with very different needs, ideologies and approaches. Weakened by both external and internal forces, OPEC ceased to be much of a force after the early 1980s. Indeed, by the late 1990s "it was on its deathbed," says Fadel Gheit, the Wall Street oil analyst. With the demise of OPEC anticipated, a key piece in the international oil picture appeared to be about to fall into place exactly as Washington had hoped.

But then, to the astonishment of many, OPEC suddenly began to revive. In 2000 it once again become a considerable force, effectively dictating world oil prices, much to the annoyance of Washington.

How did this happen? "It was Hugo Chávez," says Gheit. "He saved OPEC."

CHAPTER 4

REVOLUTION AND ICE CREAM IN CARACAS

As his motorcade pulled up to the Iraqi border, Hugo Chávez knew he was about to cross a particularly significant line in the sand.

It was August 2000, and since his election as president of Venezuela eighteen months earlier, Chávez had been intensely focused on the task of breathing life back into OPEC, a goal that was crucial to the future well-being of his oil-rich nation. But Chávez was under no illusion that reviving OPEC would be easy. It meant, above all, restoring trust among the leaders of eleven nations that for years had been at each other's throats in one way or another. Here he was, after all, about to enter Iraq from Iran—a border crossing that would have been inconceivable throughout most of the 1980s, when these two important oil-producing nations were locked in a vicious, relentless war.

Now Chávez had taken it upon himself to play the role of conciliator, of mediator, of healer, in a desperate attempt to convince all the members of OPEC that they had too much at stake to let the organization become a spent force, that the future of all their nations depended on saving it. He'd already been making this case for months in messages and phone calls. But now he'd come in person to underline his commitment to the OPEC cause, and to try to make real personal connections with these very different leaders. His task was all the more difficult because he had no familiarity with Arab culture and spoke no Arabic, so he was obliged to make his impassioned pitch through an interpreter. Above all, he wanted to personally urge the OPEC leaders to come to the summit he had called for in Caracas the following month, where he hoped that together they would launch a new era for the organization.

Taking his message to Iraq posed particular problems. Not only was it physically difficult to get to Baghdad—there were no international flights into or out of Iraq—but going there involved tweaking the nose of Washington. If there was one theme that had clearly emerged in U.S. foreign policy in the 1990s, it was Washington's desire to isolate the Iraqi dictator. To this end, Washington had pushed for and achieved punitive UN sanctions against Iraq, and had enforced a ban on international flights. Of course, there was plenty of hostility towards Iraq from other quarters as well, particularly after Iraq's 1990 invasion of Kuwait. But it was Washington above all that had orchestrated Iraq's exclusion from the world community, insisting on maintaining the severest sanctions even when other UN Security Council members were willing to ease up, in order to reduce hardship within the country. And it was Washington that had pressured foreign leaders to stay out of Iraq, keeping Saddam ostracized. For a decade, not a single head of state had visited Baghdad. That was exactly the way Washington wanted things—the uncooperative, hostile Saddam effectively caged off from the world. Now Chávez was about to make a small hole in the walls of that cage. And by stepping over the Iran–Iraq border, he was crossing a line that Washington had made clear it didn't want crossed.

Chávez's motorcade had barely entered Iraq, at a spot 180 kilometres northeast of Baghdad, when a State Department spokesman in Washington expressed U.S. disapproval, saying it was "particularly galling" that Chávez, a democratically elected leader, had broken the Iraqi dictator's isolation from the international community. Once inside the country, Chávez was whisked by Iraqi military helicopter to Baghdad, where he held talks with Saddam and toured the capital in the Iraqi leader's private car. At a press conference later that day, Chávez dismissed U.S. objections to his visit. "We regret and denounce the interference in our internal affairs. We do not and will not accept it."

Chávez then resumed his oil shuttle mission, which included a visit to another state on Washington's list of pariahs—Libya. Chávez knew he was courting problems with Washington, but he also knew that any hope of resolving the suspicions and hostilities that had long bedevilled OPEC depended on this sort of personal diplomacy. And he was determined to overcome any obstacles. The fate of his own ambitious agenda of reforms for Venezuela depended on significantly higher oil revenues, and that required a stronger, united OPEC.

The virtual disintegration of OPEC in the late 1990s had allowed oil prices to plunge to below ten dollars a barrel, depriving Venezuela, and the other OPEC nations, of precious revenue. It was the perennial problem OPEC faced: while all its members wanted to maintain oil prices at a high level, they also each wanted a large share of the market. The only way to keep prices high, however, was to restrict supply. Hence the need for quotas, with each country required to stay within its limit. But since OPEC lacked an effective policing and disciplinary procedure, countries tended to cheat, producing more oil than their quota allowed. The result was a flood of oil on the market, which brought down the price.

All this was clearly self-defeating from the producing countries' point of view. Still, the temptation to cheat was great, partly because each country assumed the others would as well. For one thing, many OPEC countries were poor, with large populations. Their economic needs were therefore considerable, and oil was usually their main source of income, often their only source of income. So boosting their production was a quick way to earn desperately needed revenue—until other OPEC countries did the same and the price fell.

What was clearly needed was some sort of internal discipline. That's hard to come by in any organization, of course, and particularly in one involving sovereign nations with so many differences and long-standing grievances. In addition to the wars waged between OPEC members, there were the huge cultural and ideological differences between, for instance, an arch-conservative, pro-American state like Saudi Arabia and a radical, anti-American state like Libya. Furthermore, there were huge variations in the needs and assets of the various OPEC countries. Kuwait and the United Arab Emirates had relatively small populations and huge oil reserves, while Indonesia and Nigeria had huge populations and relatively small reserves. These sorts of differences led to different strategies for maximizing oil revenues, making internal cohesion in OPEC difficult and quota-cheating a constant problem.

In fact, when it came to quota-cheating, Venezuela had been one of the worst offenders, constantly provoking the wrath of Saudi Arabia, the effective master of OPEC in the 1980s and 1990s. But Chávez was determined to change Venezuela's delinquent behaviour. And one of the key stops on his whirlwind tour in August 2000 was the Saudi capital of Riyadh. Despite its massive wealth, Saudi Arabia had been suffering badly from depressed oil prices. Ordinary Saudis had become accustomed to a high standard of living, and the low oil prices of the late nineties had left the kingdom unable to properly fund public programs, including health care, that its people had come to expect. Chávez's message—that it was essential OPEC pull itself together—resonated strongly with the Saudi rulers.

Chávez's idea was simple. OPEC needed to establish a "price band," in which the price of a barrel of oil would be allowed to go no lower than $22 and no higher than $28. In order to maintain this band, OPEC nations would have to agree to automatically cut production whenever the price of oil sank below $22 a barrel and automatically boost production whenever the price rose above $28. Each nation would have to accept the price band and the automatic production cuts or increases required to maintain it.

The idea had the potential to deliver enormous benefits to all OPEC members by significantly increasing the price of oil from the depressed 1999 level. Even if nations were obliged to cut back the amount they produced, they would be getting so much more per barrel that they would likely be better off. And they'd be able to save more of their reserves for the future. OPEC could even argue that the consuming nations would ultimately benefit, since the price wouldn't go above $28 a barrel, thereby preventing the kind of sharp price spikes that provoked recessions. While it sounded good on paper—at least to OPEC members—the real question was whether the countries could overcome their long-standing mutual mistrust. That would ultimately be determined at the summit Chávez hoped to convene in Caracas.

His shuttle diplomacy paid off. By the end of his whirlwind trip, every nation had agreed to come.

There hadn't been a summit of OPEC heads of state in twenty-five years, but in late September 2000 such a gathering took place in Caracas. The badly ailing King Fahd of Saudi Arabia didn't make it, nor did Saddam Hussein, who for security reasons no longer left Iraq. But both leaders sent deputies with full decision-making powers, operating as de facto heads of state. The negotiating sessions went on for three days and were often tense, particularly during the long, gruelling final day. But the summit ended with essentially the agreement Chávez had sought. After tottering on the verge of extinction only a year earlier, OPEC was very much alive.

Chávez had played the key role in the resuscitation. It was just one of the ways in which the radical, nationalistic Third World leader was shaking up the international oil scene—and in the process making himself *persona non grata* in Washington.

———

Despite previous disappointments, I always retain the faint hope that the hotel I have booked will turn out to be a delightful little place that perfectly captures the local charm of my foreign desti-nation—in this case Venezuela. So it is with a particular thud in my heart that I arrive at the front door of a Holiday Inn-style hotel, cut off from downtown Caracas by a network of expressway-like roads and attached to a sprawling indoor shopping centre, which comes about as close to capturing the sights and sounds of Venezuela as, say, the West Edmonton Mall.

Looking on the bright side, I figure that such a dreary North American-style complex will at least be safe. After all, it's located in a well-to-do part of town, far from the desperate poverty of the shantytowns that surround the city. But it turns out that it's the well-to-do parts of Caracas that are actually the dangerous ones. The week before my arrival in March 2004, anti-government agitators in an exclusive neighbourhood near the hotel had started bonfires and set up barricades in the streets. It's the rich who are agitating in Venezuela these days, hell-bent on overthrowing their left-leaning, democratically-elected president.

Hugo Chavez comes from modest roots (the son of low-paid schoolteachers), and he swept to power in 1999 with massive sup-port from the nation's poor, who make up well over half of the country's population of 23 million. Since then, he's pushed the well-to-do in Venezuela out of their privileged positions running the country and left them raging from the sidelines. They now seem to spend much of their time plotting his downfall. A key tool has been their ownership of local TV stations, which, along with exhaustive coverage of Hollywood celebrities, seem to focus tire-lessly on the evildoings of the president.

Back in April 2002, a faction of the elite, led by the head of the local chamber of commerce, organized a coup in which Chavez was taken prisoner. But no sooner had the wealthy elite turned up at the palace in their finery to celebrate than tens of thousands of poor people clogged the streets, demanding their president back. Much of the army ended up siding with Chavez, who had risen through army ranks himself. Within forty-eight hours, Chavez was back in charge, more than ever the hero of the poor and the scourge of the rich.

For a guy who remains in the crosshairs of the country's powerful elite, Chavez seems remarkably cool. There appeared to be little security for instance at a public event in Caracas where Chavez was speaking on International Women's Day. The place was jammed with people, many of whom looked downright poor; next to me in line was a woman in a ragged peasant dress and bare feet, with a several hungry-looking children in tow. I explained to the security guards that I was a foreign journalist, and was asked for a press pass. Since I didn't have one, I flashed the most official-looking thing I could find in my wallet—a University of Toronto library card. Several guards scrutinized it carefully and then, without even checking for overdue books, waved me through. I suspect that ID from my local video store would also have worked.

Security is a little tighter at the presidential palace. Once inside the iron gates, I discover a lovely colonial-style mansion, all gleaming white, with a courtyard of palm trees and flowers, and rooms with intricately carved, gold-trimmed furniture and elaborate chandeliers. Crisply dressed waiters serve guests ice-cold drinks from silver trays. I'm led through a stately room where historic dignitaries stare down from heavy portraits high on the walls. A sudden turn and we're heading up an unadorned back stairwell to a small, rather ordinary conference room where Chavez rises to greet me, wearing casual garb and Hush Puppies.

Venezuelan presidents are typically distinguished-looking white men, descendants of the Spanish conquerors. But Chavez is part black, part Indian—a descendant of the conquered people whose traditional place at the palace would more likely be in the pantry. This seems to irk the elite, who have been known to refer to their president as "the monkey." What irritates them more are his efforts to redirect the country's substantial oil revenues from their hands to the general welfare, and his determination to steer Venezuela away from the package of neoliberal policies aggressively peddled by Washington.

This sort of behaviour could land Chavez in serious trouble. The Bush administration considers Venezuela under Chavez to be, if not a rogue state, then at least a mighty annoying one.

Chavez is convinced that Washington was involved in the April 2002 coup against him. "Washington applauded," he noted. "The American ambassador came here [to the palace] and supported the coup."

The U.S. did support the coup—at least initially. Charles Shapiro, the U.S. ambassador, met with the coup leaders after they took control of the palace and was quick to accept their assertion that Chavez had "resigned" and that they constituted the legitimate new government of Venezuela. The next day, after Latin American leaders strongly condemned the coup, U.S. Secretary of State Colin Powell spoke out against it.

Evidence has since emerged which suggests that the U.S. may have helped orchestrate the coup: anti-Chavez elements within Venezuela received extensive funding from Washington, and a military attaché to the U.S. embassy in Caracas, Lieutenant Colonel James Rodger, is reported to have been in the Army Command buildings where rebel military factions were contributing logistical support during the coup, and to have advised the rebelling generals. Certainly, some key players in the opposition appear to have strong ties to high-level people in Washington: Gustavo Cisneros, a fabulously wealthy Venezuelan businessman, is reported to be a personal friend of George Bush Sr., and the top Venezuelan oil official from the pre-Chavez days, Luis Giusti, has served as an energy adviser to the administration of George W. Bush.

Chavez is apparently not cowed by the prospect that he could end up facing the same fate as Haiti's Jean-Bertrand Aristide, another democratically elected president who offended both the local elite and Washington. (Finding his palace surrounded by armed thugs in early 2004, Aristide discovered he had little choice but to submit his "resignation" to U.S. army personnel. Aristide says he signed the letter under threat of force and was effectively kidnapped. One question: why would there be a need for a signed letter of resignation, if the resignation were voluntary?)

"The Bush administration has been invaded by madness," Chavez tells me, echoing a sentiment heard in just about every corner of the world, but rarely so openly expressed by a national leader. But, then, Chavez has a reputation for being something of a free spirit. Watching him at the International Women's Day event, as he heartily belted out popular love songs for the boisterous crowd, I was struck by this unique presidential package—military man, orator, class warrior, crooner.

My interview with him stretched to two and a half hours. Towards the end, an official came in to point out that pressing matters of state awaited his attention. Chavez responded with a request that they bring us some ice cream. Shortly, chocolate sundaes, with cherries on top, were brought in. In between mouthfuls, we talked about the invasion of Iraq—and its relevance to Venezuela.

Chavez believes the Bush administration invaded Iraq to get control of its oil and that it harbours similar ambitions towards Venezuela's oil. "It's the people around Bush, the great oil interests," Chavez said. Then in a sing-song voice, he gently chanted: "Cheney, Cheney, Cheney."

Chávez wants to realign Venezuela more closely with the developing nations of the South, particularly those in Latin America, but also in Africa and Asia. "There is never only one road," he says. "But I believe the best of roads for the Southern countries to take is to be integrated together, the South with the South." He talks about the possibility of establishing a strong infrastructure for the Southern countries: a Bank of the South, a University of the South, a Secretariat of the South, a TV channel of the South. "The North comes straight to us," he says, suggesting that the southern countries need to collectively counter this domination by the North so they can create "an alternative model to globalization."

With this sort of thinking, Chávez has emerged as one of the more radical leaders inside the G-19, an alliance of Southern countries trying to counter the power of the industrialized nations of the North. It has also led him to develop close ties with Cuban dictator Fidel Castro, whom he describes as the "leader of the South with the greatest experience." Chávez's dealings with Castro—Venezuela now provides oil to Cuba in exchange for Cuba sending doctors to Venezuela—have been another source of friction with Washington and with Venezuela's own elite. Chávez's real source of revolutionary inspiration, however, is Simón Bolívar, the nineteenth-century anti-colonial leader who fought the Spanish in an attempt to bring about a united, post-colonial Latin America. (Since Chávez came to power, major roads and buildings in Venezuela have been renamed in honour of Bolívar; indeed, the country has been renamed the Bolívarian Republic of Venezuela.) Chávez argues that North and South America share a common past of being under European colonial domination. "That was tragedy number one." But while the North has shaken its colonial past, he argues that the South is mired in a new form of colonial domination—by the North. "That is tragedy number two . . . We continue to be sunk in tragedy."

In his efforts to foster unity—and defiance—among the countries of the South, Chávez has held up OPEC as a useful model, even a beacon of hope. "I have given the example of OPEC," he says, noting how unity has enabled OPEC nations to stand up to the industrialized countries. The notion of OPEC as a model for challenging northern power helps explain Washington's continuing hostility to the organization, which might otherwise seem puzzling. After all, OPEC's central aim is to maintain high oil prices. So, logically, the big oil companies—and their supporters inside the White House—should be happy to let OPEC run the international oil scene. But they aren't, and the reason appears to be that they don't like such power residing in a handful of Third World nations. They want it to reside instead in Washington, and by extension in the international oil companies with which the Bush administration in particular is so closely allied. As well, Washington sometimes feels the need to bring oil prices down, in order to please other domestic constituencies— something that OPEC would resist. Therefore, the more Washington can weaken OPEC, the more flexibility it will have to keep oil prices where it wants them, and thereby control an important lever in the global economy. "Ronald Reagan threatened to bring OPEC to its knees," says Chávez. "The Americans have tried to satanize OPEC, saying it was the cause of the economic crisis [of the 1970s and early 1980s]. They've dedicated themselves to weakening the unity of OPEC."

There have been other attempts to organize Third World producing nations into effective associations—around commodities like coffee, cocoa and copper—but such attempts have all failed, leaving Western multinational corporations dominating these industries and producer nations receiving almost nothing. Oil has been the sole breakthrough commodity, the Third World's one foot in the door. "None of the others got off the ground," notes economist Michael Tanzer, who advises Third World governments. "OPEC alone is an island of relative power." As such, OPEC serves as an inspiring model of how unity and coordinated actions can achieve results for the developing world, even against overwhelming odds. Of course, oil is a special case because of its enormous value. But even so, OPEC has had to overcome great difficulties—difficulties Chávez believes Washington exacerbated with attempts to weaken OPEC unity, even to convince OPEC nations to break from the organization.

Chávez believes that, like Ronald Reagan, top officials in the Bush administration would like to bring OPEC to its knees. "They see OPEC as an obstacle for them to impose their will on world oil policies, through the transnational companies." Tanzer agrees with Chávez's assessment. "Washington is still resisting OPEC . . . It's the political power of OPEC that worries them."

Chávez also believes that Venezuela's role in helping reunite OPEC has inflamed Washington against his government. "We have been a government absolutely independent in designing our oil policies," he says. "We have not accepted pressures of any type to separate ourselves from OPEC . . . They would like to manipulate OPEC, and they know the role Venezuela has played in OPEC, and that's made us a target."

———

Shortly after he was first elected, Chavez held a referendum on whether a new constitution should be written, and won support from more than 70 percent of the electorate. A special assembly was then elected to draft a new constitution. There were extensive consultations with the broader public, including those lower down the income scale who aren't normally consulted on these sorts of things. The result was a radical document that includes, among other things, strong protections for women's equality, rights for indigenous people and a ban on the privatization of the nation's oil. Written in plain, everyday language, the new constitution is available in a tiny, pocket-size version that is sold by street vendors in Caracas. The constitution was one of the first things targeted by the armed faction that stormed into the presidential palace in April 2002. Coup leader Pedro Carmona, the head of the Venezuelan chamber of commerce, announced that he was dissolving the Supreme Court and the elected Parliament, and that the new constitution was null and void.

Recalling that night, Chavez says: "For the first time in our history, we got control of the oil. But the tyrant declared the constitution invalid."

Getting control of the oil was no easy task. Oil was first discovered in Latin America at the end of the nineteenth century and it quickly attracted the interest of American oil companies.

"The Americans have always been involved with Venezuela, particularly with oil," says Frank Bracho, a former Venezuelan diplomat and author of *Petroleum and Globalization*. As early as 1903, an attempted coup financed by a U.S. oil firm, the New York and Bermudez Company, led to deeply strained relations between Washington and Caracas, and soon afterwards to the severing of all diplomatic ties. A more effective coup attempt, again financed with U.S. oil company money but this time backed up by a U.S. naval fleet, brought General Juan Vicente Gomez to power in 1908. Oil production was soon underway and concessions were parcelled out to a small group of Gomez's family and friends, who then sold the rights to foreign oil companies, which were taxed at a very minimal rate. Thus began a close collaboration between the country's elite and foreign oil interests—a collaboration that was to become more entrenched over the decades, with the country's immense oil wealth largely disappearing into private hands, both at home and abroad. For the millions living in the city slums and throughout the countryside, the oil might as well not have existed.

There were attempts to change this state of affairs, particularly in the late 1950s and the 1960s, when a visionary Venezuelan politician, Juan Pablo Pérez Alfonzo, emerged onto the political scene and became one of the founding fathers of OPEC—a story we'll return to in Chapter 8. Pérez Alfonzo also tried to assert public control over Venezuela's oil resources, with minimal success. Eventually, in the mid-1970s, with most oil-producing countries nationalizing their oil industries, Venezuela did so as well, taking over concessions that had been run by Exxon, Mobil and Shell. Venezuela's oil would henceforth be handled by the state-owned oil company, PDVSA.

But despite the public ownership, little changed. PDVSA ended up being run by Venezuelan managers who had worked for the U.S. oil companies, and apparently preferred answering to U.S. bosses than to the Venezuelan government. In fact, the PDVSA managers, who had been educated in America and spoke fluent English, were deeply oriented towards the U.S. "They were very close to American culture," says Bracho. "Americans said they loved to work with PDVSA guys 'because they're so much like us.'"

There was a deliberate policy of keeping PDVSA autonomous, but the effect of that autonomy over time was to turn the company into a force of its own, "a state within a state," as Venezuelans often describe it. Attempts by various Venezuelan political leaders to bring the state-owned company truly under state control were successfully rebuffed. PDVSA developed a large, top-heavy bureaucracy, which became widely known for its corruption. Managers awarded themselves enormous salaries while transferring vast sums of money to private companies they controlled outside the country. Rafael Ramirez, minister of energy in the Chávez government, charges that some $10 billion was transferred out of the country by corrupt PDVSA officials, many of whom now live in exiled luxury in Florida, alongside wealthy Cuban exiles with whom they make common political cause.

Meanwhile, the PDVSA bureaucracy just continued to grow, as managers kept hiring family members and friends for cushy jobs that involved high pay and little work. Stories of wastefulness and nepotism inside the oil behemoth became legendary throughout Venezuela. All salaries were paid in U.S. dollars, and there were generous perks. Even employees in low-level administrative or personnel jobs—often hired because of family connections—routinely received thousands of dollars in bonuses. PDVSA became a world of its own, a privileged, self-perpetuating dynasty in the midst of a land of poverty, feeding off the country's fabulous oil wealth.

The independence of the company and those who ran it was symbolized by the fact that PDVSA was housed in a number of posh modern office towers while the government oil ministry, which ostensibly controlled it, was located in humble, somewhat dilapidated quarters. In practice, PDVSA didn't really answer to the government at all. A former PDVSA president, Brigadier General Guaicaipuro Lameda, dismissed the notion that the company should be under government control. The company, he argued, had "a twenty-seven-year history of being efficiently run as a profit-making company that pays dividends to its shareholder, the state. It shouldn't be delegated to the inferior status of being a mere appendage of the oil ministry, subject to the president's interference." Chávez says that the state-owned company didn't even provide the state with a proper accounting. "They robbed us," he says. "There was double counting of the money. The national treasury received very little money."

With the PDVSA management largely independent of the Venezuelan government, it continued to grow ever closer to foreign-based oil interests. By the late 1980s, Luis Giusti, then president of PDVSA, decided to reopen Venezuela to foreign oil companies, granting them access to oil fields that were ostensibly marginal and not particularly productive. Some weren't all that marginal, however. Exxon and BP returned to the country with extremely favourable contracts to develop what turned out to be highly productive oil fields.

The effect was to boost Venezuela's oil production above its OPEC quota, incurring the wrath of Saudi Arabia and weakening the unity of the cartel. Of course, weakening that unity was something to be encouraged as far as Washington was concerned. So it was also fine with PDVSA management, which identified with Washington and the multinational oil companies rather than with the Third World nations of OPEC. Increasingly, PDVSA attacked OPEC in its public statements, calling the organization a hindrance to Venezuelan oil development and rejecting the quotas as unduly restrictive. By the late 1990s, PDVSA had already allowed Venezuelan production to rise to 3.5 million barrels a day—well above its OPEC quota—and was talking openly about allowing production to go as high as five million barrels a day. All this contributed to the near-death of OPEC. Meanwhile, Giusti was apparently contemplating another bold move. "He was preparing the privatization of PDVSA," insists Chávez.

The election of the Chávez government put a spanner in the plans of PDVSA. The new constitution, which came into effect in 1999, specified that the nation's oil was not to be privatized. Furthermore, Chávez began to demand more financial information from the company, and he put three of his close associates on the board—appointees who were quickly denounced as "incompetent" by the company management. With these moves Chávez was beginning to bring the company under government control, to make it the very "appendage of the oil ministry" the PDVSA management had long resisted becoming. The government further infuriated the oil elite by passing a controversial law raising government royalties on oil produced by both PDVSA and foreign companies to 30 percent from 16.7 percent. The law also required that the state oil company have a majority stake in all joint ventures with foreign oil companies.

When the coup leaders briefly took power in April 2002, it was clear that one of their goals was to regain full control over PDVSA and oil policy. In their forty-eight hours in power, they found time not only to declare the new constitution invalid but to signal their intention to cancel oil shipments to Cuba and to resume an anti-OPEC stance. "One of their first declarations," says Chávez, "was that Venezuela would separate from OPEC."

The coup failed, but the fierce opposition towards the government within PDVSA only intensified, and it led to a showdown later that year. In December, PDVSA shut down all oil operations in the country in a direct challenge to the government. Although there was some support for the shutdown from the leaders of the oil workers' union, the "strike" was effectively run by the senior management, with support from thousands of white-collar workers in the company's oversized bureaucracy. Workers in the oil fields found themselves locked out. "It was a tremendous showdown," notes Frank Bracho. "Whoever controlled the oil, that side would have the political power."

The business community quickly moved to offer its full support to the PDVSA management, declaring a "general strike" and closing major stores and offices, in an effort to force Chávez to resign. The strike also received support from some U.S. companies operating in Venezuela, including McDonald's and FedEx, which closed their operations in solidarity with the elite. The wealthier eastern portion of Caracas was largely shut down. Meanwhile, in the central and western parts of the city, there was little apparent support for the "general strike," according to Mark Weisbrot from the Center for Economic and Policy Research, who was in Caracas at the time. He reported that "outside the wealthier areas of eastern Caracas, businesses are open and streets are crowded with shoppers. Life appears normal. This is clearly a national strike of the privileged, and most of the country has not joined it." With the big grocery stores closed, the government set up giant tents for improvised markets where farmers could sell their produce directly to city-dwellers.

Meanwhile, the U.S. media, which draw much of their footage of Venezuela from the private TV stations there, portrayed the strike very sympathetically. When the Venezuelan government sent the military to recover oil tankers seized by striking captains, the U.S. media accused Chávez of using "dictatorial powers." The absurdity of this charge was captured by Weisbrot: "[W]hat would happen to people who hijacked an oil tanker from ExxonMobil in the United States? They would be facing a trial and a long prison sentence." Weisbrot also noted the hypocrisy of the American media's obvious sympathy for the Venezuelan "strikers" when similar actions by ordinary American workers wouldn't be tolerated. "[T]he employees of the state-owned oil company—mostly managers and executives—are trying to cripple the [Venezuelan] economy, which is heavily dependent on oil exports, in order to overthrow the government. In the United States, even private sector workers do not have the legal right to strike for political demands, and certainly not for the president's resignation. In the United States, courts would issue injunctions against the strike, the treasuries of participating unions would be seized, and the leaders would be arrested."

The shutdown of oil operations by the strikers had potentially destructive long-term implications for the country's oil resources. The flow of oil can't abruptly be cut off without damaging the oil field. Adjusting the flow is a tricky operation, and the striking managers in many cases removed vital computer discs containing the technical data needed to carry out the operation safely and maintain the right pressure in the ground to prevent long-term damage. The government narrowly averted disaster only with emergency help from retired Venezuelan oil workers and oil experts flown in from Brazil and Trinidad. Together they were able to get the oil running again.

After an extremely tense two-month showdown between the PDVSA management and the Chávez government, the government prevailed. Its victory over the company's rebellious managers was complete. Some eighteen thousand employees who had gone on strike were fired. Soon afterwards, in a move full of symbolism, the government oil ministry was relocated from its former humble quarters into one of the massive office towers that had served as headquarters for PDVSA's senior mandarins. And another huge PDVSA management building was transformed into a free university for the poor, with former PDVSA vehicles used to bus students every day from their barrios.

———

A new source of tension between OPEC and Washington was the sudden spike in oil prices in March 2004. With the U.S. presidential election looming, rising domestic gasoline prices suddenly flared up as an issue, with OPEC taking the heat. It had orchestrated the increase, pushing the price of oil above its own self-imposed limit of $28 a barrel. This was a deliberate attempt by OPEC to recover some of the money its members lost with the steady decline in the value of the U.S. dollar. With OPEC oil priced in U.S. dollars, member nations had watched their buying power decline as the dollar slid in value. To compensate, OPEC decided to let the price of oil rise above the $28 cut-off, into the mid-$30 range. Some factions within OPEC even argued for abandoning the U.S. dollar altogether and adopting the European currency instead—a move, it is widely assumed, that would displease Washington. (Indeed, Saddam Hussein had taken this provocative course, opting to price Iraqi oil in European currency.)

In the months that followed, oil prices kept climbing, and seemed stuck above $50 a barrel in the spring of 2005. This time, however, the higher prices were no longer being orchestrated by OPEC, but were related to broader supply problems due to rising demand, the deteriorating situation in the Middle East and the fast-approaching peak in world oil production. Although the OPEC countries weren't the cause, they were certainly—along with the major oil companies—the beneficiaries.

Meanwhile, some environmentalists argue that high oil prices may be necessary for the broader goal of curbing the world's oil consumption, which is showing no signs of slowing down despite the looming threat of climate change. Needless to say, high oil prices aren't the ideal solution to the oil consumption problem, since they place such a heavy burden on undeveloped countries and on low-income consumers in the industrialized world. For the time being, however, high prices seem to be the only hope for constraining the world's self-indulgent oil consumption.

Lots of better, more equitable solutions exist, but that doesn't mean Western governments have the slightest interest in implementing them.

CHAPTER 5

FROM COFFINS TO WORLD DESTRUCTION
THE TALE OF THE SUV

Faced with a particularly bleak Christmas Eve—after his beautiful wife had left him, taking their only child with her—Svante August Arrhenius did what any distraught man might be expected to do: he stayed up most of the night working on mathematical equations. Even under ordinary conditions Arrhenius, a nineteenth-century Swedish chemist, was an obsessive character, given to all-night bouts of scientific inspiration. (Eleven years earlier, in 1883, he had lain awake all night thinking about how good conductors of electricity decompose into atoms carrying chemical charges when submerged in water.) Still, even by Arrhenius's standards, the decision to spend all of Christmas Eve 1894 working on the calculations was an odd one, since he considered the calculations very dull work. Even odder was the fact that, almost a year later, he continued to work almost around the clock on the ever-growing and -expanding set of calculations, which he had come to regard as the "most tedious" of his life. As he wrote to a friend, "It is unbelievable that so trifling a matter has cost me a year."

The trifling matter had to do with the question of how the temperature of the earth's surface is influenced by gases in the atmosphere—admittedly not a question that occupies most people's minds on Christmas Eve. It wasn't that Arrhenius was particularly drawn to questions about the earth's surface or the role of gases in the atmosphere, but rather that it represented a puzzle, a piece of unsolved science that needed an answer. Others had pondered a related question before—how did the earth stay warm?—and had figured out important aspects of the mystery. But in his dreary, year-long set of calculations, Arrhenius eventually came up with a key piece of the puzzle, which, although widely ignored at the time, would eventually grab the world's attention and point to a particularly worrisome future.

Of course, it might seem intuitively obvious that the earth is warm because the sun warms it. But as far back as the 1820s scientists had figured out that this was only part of the story. The sun's rays do indeed shine down on the earth, warming it, but, as the French scientist Jean-Baptiste-Joseph Fourier figured out, those warm rays from the sun actually bounce back when they hit the earth's oceans and land masses. And that bounced-back heat would simply escape back into the universe, leaving the earth unbearably cold, except for one thing: the invisible layer of gases—particularly carbon dioxide—that hovers just above the earth. Those gases absorb part of the bouncing-back heat and trap it—just as a greenhouse does—and in the process make life on earth possible for humans, among others.

All this is very reassuring, suggesting that some great power (God, Nature or some sort of Supreme Being beyond mere human comprehension) has carefully thought through things and come up with a workable scheme, without which none of us would be here contemplating any of this. It would be nice simply to leave it at that, to just assume that the system works, that there's some giant self-regulating thermometer out there taking care of things while we get on with our lives. But the story gets more complicated. Building on Fourier's work, other nineteenth-century scientists went on to elaborate about how the earth's heating system works. An Irish mathematician and engineer, John Tyndall, came up with a way of measuring how effective carbon dioxide was at absorbing and trapping the sun's rays, and speculated in a scientific paper in 1861 that a drop in the earth's carbon dioxide levels could lead to another ice age. About twenty years later, U.S. astronomer Samuel P. Langley theorized that mountains are colder at the top because the layer of gases up there is thinner, allowing more of the sun's heat to escape back into outer space. These observations and speculations pointed to the possibility that the earth's temperature could change if the amount of carbon dioxide in the atmosphere changed, but nobody seemed to see any big changes coming on the carbon dioxide front. Nobody, that is, except Arrhenius.

Up until this point, it was assumed that carbon dioxide levels in the atmosphere were determined largely by natural processes involving plants and rocks. It was known that the burning of coal released some carbon dioxide too, but the fact that industrial factories were burning large amounts of coal had not been considered significant in terms of the earth's overall carbon dioxide levels. Arrhenius, however, was struck by the extent of the coal-burning that was going on in the Industrial Revolution and considered that it might be having a greater impact than was appreciated. He had travelled widely through northern Europe, had witnessed first-hand how factories were transforming the landscape and the air above cities. All this had obvious side effects, like making the air dirtier and breathing more difficult. But Arrhenius also wondered what it might be doing to that delicate earth-heating system Fourier had described. It was this question that was running through his mind as he sat down to do some calculations on that fateful Christmas Eve.

After boring himself almost senseless with endless equations, he became the first person to draw the conclusion that humans would alter the world's climate with their relentless burning of fossil fuels. It was a stunning conclusion that would eventually force the world to reassess its own behaviour, unleashing an enormous political struggle between those willing to make changes and a small group refusing to. But at the time, back in the 1890s, Arrhenius's finding was of no momentous interest even to those inside the scientific community, and certainly of no interest whatsoever to those outside it. In fact, his discovery of the crucial role human actions can play in altering the earth's climate remained obscure, even after Arrhenius won the Nobel Prize for chemistry in 1903 (for work unrelated to the earth's temperature). Arrhenius himself seemed fairly unimpressed by his own scientific breakthrough on climate—largely because, despite the time he invested in the boring calculations, he made a critical error.

No wonder he considered his breakthrough of fairly middling interest: his assumptions led him to conclude that it would be an awfully long time before the burning of fossil fuels caused carbon dioxide levels to double, at which point he figured there would start to be an impact on the world's climate. He estimated that the doubling wouldn't occur until about the year 5000—an outrageously long time in the future, hardly something to get worked up about. But he'd reached that comforting conclusion by assuming the world would continue to consume fossil fuels (coal, oil, gas) at the same rate it was consuming them in the 1890s. That turned out to be dead wrong. Over the next hundred years, industrial production increased fifty-fold, and the oil-powered car became the basic means of transportation throughout most of the world. All this rendered Arrhenius's calculations meaningless. Given the vastly higher fossil fuel consumption of the last century, it now seems that his estimate was off by about 2,950 years. The crucial doubling point is expected to happen some time midway through this century; perhaps much sooner.

Sort of puts a different spin on things.

―――

In the early 1980s, the world's scientific community came together with a sense of urgency to tackle a serious problem: the depletion of the earth's ozone layer, another problem related to gases in the atmosphere, although different from global warming. The ozone depletion had been identified in the early 1970s, and scientists had figured out that it had something to do with the production of certain chemical products, such as the aerosol used in spray cans and certain kinds of refrigerants and coolants. They also figured out that the depletion was potentially serious for humans, leaving us more vulnerable to the sun's ultraviolet rays, which are normally blocked by the layer of ozone above the earth. That layer had developed a big hole. With remarkable speed, the scientists determined what steps governments would have to take to repair the rapidly growing hole.

If there was one person central to solving the ozone problem, it was Robert Watson. After graduating from the University of London with a Ph.D. in atmospheric chemistry in 1973, Watson had fully expected to spend his life immersed in the highly academic study of the chemical reactions of chlorine, bromine and fluorine. But the following year, just as Watson was settling into post-doctoral research at the University of California with plans for a lifetime of study, two American scientists reported findings that indicated the world's ozone layer was rapidly being depleted. The damage was reportedly being done by the very chemicals Watson specialized in. Watson began working on research projects related to the ozone problem, which were funded by a larger research effort conducted by NASA, the U.S. space agency. Suddenly he was caught up in trying to apply scientific knowledge to better the world; his days as a purely academic scientist were over.

Watson proved to have a keen political sense of what was involved in tackling a problem this huge. While dozens of scientists contributed to the growing knowledge about the ozone layer depletion, it was Watson who grasped something crucial early on: the real obstacles lay outside the realm of science.

Powerful industrial interests stood to lose financially from the changes that would be required, and counteracting their influence would be the central problem. What was needed was a highly credible process for assessing the scientific data so that the data could not be dismissed by big companies. Watson thus launched something ambitious—more ambitious than anything that had ever been tried—in the interest of getting the world to pay attention to science: a massive international process in which scientists all over the world who specialized in this field were consulted. Hundreds of scientists were asked to critically review the data, and to do so quickly and comprehensively.

What emerged from the process was perhaps surprising: a stunning degree of co-operation and consensus about the seriousness of the problem and the nature of the solution. The sheer strength of the response had the effect of undermining the corporate efforts to derail the process. DuPont, General Electric and a few other industrial giants had mounted campaigns to suggest ozone depletion wasn't a problem, even found a few scientists willing to support their case. But the corporate campaigns soon petered out, overwhelmed by the strength of the evidence and the sheer length of the line-up of scientists on the other side. Eventually, the once-resistant companies got on board and actively contributed to the search for a solution. By the late 1980s an international treaty—worked out at a Montreal meeting and known as the Montreal Protocol—had been drawn up. Within a few years, ninety-three countries, including the United States, had signed, committing themselves to phasing out the offending chemicals by the end of the decade. The world would have to learn to live without aerosol spray deodorants. Somehow we'd get by.

The world's first truly global attempt to stop damage to the earth's ecosystem had been a smash success. Perhaps "globalization" didn't just have to be about forcing countries to open their markets to foreign ownership; it could also be about co-operating internationally in the interests of saving the earth. It was almost enough to create confidence that opposing factions would bury their differences when something as important as the future of the planet was at stake.

That confidence would have been misplaced.

———

For all the careful thinking of Svante August Arrhenius and others who had contemplated how the sun's rays are trapped for the benefit of the earth, the world hadn't really given much thought to the problem of "global warming" before it suddenly loomed large on the political radar in the mid-1980s. An international scientific conference in Toronto in June 1988 helped draw world attention to the global warming threat. The conference, at which Prime Minister Brian Mulroney was the opening speaker, ended with a consensus statement that "humanity is conducting an unintended, uncontrolled, globally pervasive experiment whose untimate consequences are second only to global nuclear war." With evidence suggesting that the phenomenon was already underway—melting icebergs, rising sea levels, unexplained weather extremes—the United Nations and the World Meteorological Organization moved quickly, setting up an independent body, called the Intergovernmental Panel on Climate Change (IPCC), to review scientific understanding of the problem. Robert Watson, fresh from his effective handling of the ozone review process and now a senior scientist at NASA, was enlisted as a key figure in the new UN body, and seven years later became its chairman.

The quick establishment of the IPCC showed how keenly the world community wanted answers and solutions. Everything seemed to be unfolding pretty much as it should, just as it had with the ozone layer: a serious problem had been discovered, and the world's nations had quickly teamed up with the appropriate scientific body to begin the process of discovering the true nature of the problem and how to tackle it. So far, so good.

From his experience with the ozone process Watson understood that the resistance to tackling global warming would come from industry. He also sensed that it would be much fiercer this time around. It was one thing to interfere with markets for spray deodorants and refrigerator cooling chemicals, which, while profitable for DuPont and General Electric, were not central to their business empires; interfering with markets for oil, gas and coal was an entirely different matter. Watson realized he would be treading on territory that lay at the very heart of the most immensely profitable and powerful set of business interests on earth.

Watson's response was to be even more thorough and comprehensive in his scientific review than he'd been the last time. With such massive corporate financial clout lined up against him, there was no room for error. Any mistakes or missteps, he knew, would be relentlessly exploited by the fossil fuel lobby. So Watson put in place an almost unbelievably rigorous system for the preparation and peer review of scientific reports—a system that remains in place. Reports are drawn up after extensive research, which includes consultation with industry. A draft is first sent to a few experts then redrafted and sent to every relevant scientist in the world—about 2,500. After feedback from these experts, it is redrafted and sent back to them for another look.

"Without any question it's the most intense peer review system ever," Watson said in an interview at his office at the World Bank headquarters in Washington. "More than any journal or any institution, by an order of magnitude . . . I don't know how you could design a process more rigorous."

The IPCC's first assessment report, released in 1990, was a powerful statement of the problem, effectively a scientific throwing-down of a gauntlet to the world. It laid out clearly that the "greenhouse effect" was real, and that after ten thousand years without a significant change in temperature the earth's surface had been getting noticeably warmer since the beginning of the Industrial Revolution. At a UN-organized conference in Rio de Janeiro in 1992, which became known as the Earth Summit, the leaders of 154 countries—including U.S. president George Bush (Sr.) and Canadian prime minister Brian Mulroney—responded by signing a legally binding convention committing themselves to address global warming.

There had been pressure to go much further and actually establish national commitments for reductions in greenhouse gas emissions, but the fossil fuel lobby, led by Exxon, fought hard—within the IPCC process, with governments and with the public—to resist such concrete measures. The basic strategy was to promote doubt about the science, suggesting there was a great deal of uncertainty, when among scientists there was almost none about the basic facts. What little uncertainty there was appears to have been largely manufactured by Exxon. A number of industry front groups were set up with names that sounded neutral or even environmentally friendly: Global Climate Coalition (GCC), International Petroleum Industry Environmental Conservation Association (IPIECA), Global Climate Information Project (GCIP). The fact that these groups were all funded by the industry wasn't technically a secret, but media reports quoting officials from the groups often failed to mention that they were industry funded. Without knowing this, why would the public have reason to question what they were saying? Why would an ordinary citizen distrust a statement that raised doubts about the science behind global warming when the statement came from a group as apparently neutral and well-meaning as the Global Climate Information Project?

Even more effective in getting the industry's message across has been skepticism in the mouths of real live scientists. And of course there are some scientists—surprisingly few, actually, not more than a dozen or so out of thousands of scientists around the world—who are skeptical. What can be said about these skeptics? It's fair to say they are not an overly impressive group. Some receive grants from the industry or work for industry-funded organizations, others apparently don't. Some of them are legitimate experts in their fields, but their fields aren't directly relevant to global warming—again something that wouldn't be obvious to the public.

Even if some legitimate scientific uncertainty still exists, Robert Watson questions why this should lead us to the conclusion that it's best to do nothing about global warming. Uncertainty, he notes, can work in either direction: "We may be over- or underestimating the problem. Uncertainty is not a reason for inaction." The same is true, he argues, about the economic impact. While global warming doubters often suggest that the economic costs of tackling the problem would be prohibitive, Watson responds: "The economic costs of *inaction* may be prohibitive." For that matter, many economists argue that tackling global warming would not be particularly harmful to the economy. Nobel Prize–winning economists Kenneth Arrow and Robert Solow, for instance, collected signatures from more than 2,500 economists for a statement declaring: "[S]ound economic analysis shows that there are policy options that would slow climate change without harming American living standards, and these measures may in fact improve U.S. productivity in the longer run."

The effectiveness of the global warming skeptics has been enhanced by the large amount of coverage they've enjoyed in the media, out of all proportion to their numbers. The media have granted this handful of skeptics—many of them industry-funded— almost the same respect and space as the thousands of scientists around the world who participate in the extensive peer review process carried out under the auspices of the United Nations. That may sound as if the media are simply following some journalistic rule about covering both sides of a story, but is it appropriate to present the two positions as equally valid when one has so much more scientific expertise behind it? In any case, the media's apparent desire for balanced coverage is highly selective. Media outlets routinely ignore all kinds of points of view and positions on any number of subjects. (When was the last time the media gave meaningful space to a labour critique of the free trade deals, or a social activist critique of the tax system?) Yet the media have given enormous attention to global warming skeptics—attention that greatly exaggerates their importance within the scientific community, and has contributed to public confusion over a phenomenon that science actually understands quite clearly.

Of course, it's possible that this small group of skeptics is really on to something, that they're the Galileos of their day— and that the only organizations fearless enough to promote the findings of these insightful mavericks happen to be industry-funded think-tanks. (Why the rest of the world resists their ideas is not clear, but it's possible the rest of the world bears an animus against humanity.)

Another possibility is that the trillion-dollar-a-year fossil fuel industry has simply used its ample resources to create the illusion of uncertainty about something that is plainly evident to virtually every serious expert who has investigated it. And in so doing, the industry has poisoned the well of public discourse on the issue, putting comprehensive collective action—the only action that could possibly work in this case—beyond the world's reach.

While one can never be 100 percent sure, if one had to choose, the second possibility does seem more persuasive.

———

The campaign orchestrated by the fossil fuel and automotive industries in the aftermath of the Earth Summit in Rio was formidable, with tens of millions of dollars directed towards discrediting the science used to determine global warming. An industry-funded U.S. advertising blitz in the fall of 1997 alone cost $13 million.

The campaign proved effective. U.S. congressmen and senators, many of whom receive political donations from the fossil fuel industry, seemed happy to latch on to the notion of scientific uncertainty. Congressional hearings held in 1995 became a showcase for global warming deniers. Arch-conservative, pro-industry congressmen turned the House into an interrogation chamber for anyone connected to or supporting the UN's IPCC process, which was vilified as a liberal plot and a hostile UN attempt to impose world government on the United States. Confronted with a long lineup of internationally respected scientists telling them the scope of the global warming problem, congressmen responded angrily, rejecting the "scare scenarios." Even top experts from NASA, America's beloved space agency, were given the back of the congressional hand, and treated as less credible than the deniers. (Skeptic Patrick Michaels, a scientist receiving extensive funding from oil and coal interests, insisted before Congress that the consensus on climate change had been stifling "scientific free speech"—despite an almost endless and highly publicized airing of the skeptics' "stifled" views.) By May 1996, angry members of the House of Representatives Science Committee, convinced they had heard enough whining about global warming, finally took action: they voted to cut funding to the government's climate research programs, eliminating some completely. Explained congressman Dana Rohrbacher, whose subcommittee had set up the hearings: "I think that money that goes into this global warming research is really money right down a rat hole."

In the Senate, the resistance was just as fierce. In July 1997, the U.S. upper chamber unanimously passed a resolution (95–0) urging the Clinton administration to reject any international deal on reducing greenhouse gas emissions that didn't spread the burden to the Third World. This unwillingness to cut the Third World any slack over their greenhouse gas emissions was astonishing. After all, as Watson points out: "We in the industrial world caused the problem . . . We got rich burning cheap coal." (Furthermore, there was a precedent for exempting the Third World from the initial round of reductions, which was all that was being urged. In the case of the ozone layer depletion, the Montreal Protocol had exempted the Third World from the initial round of reductions because Third World countries had been only minor contributors to the problem. Under the protocol, the Third World was required to begin reductions at later specified dates. Elizabeth May, executive director of the Sierra Club of Canada, which has been actively involved in both the ozone layer and global warming issues, argues that there's no reason the same system of requiring slower action from the developing world, which worked well in the case of the ozone layer, couldn't work for global warming as well.)

Meanwhile, outside the rarefied world of Congress and the fossil fuel industry, the campaign against tackling global warming was getting lonelier and lonelier. The Clinton administration was taking a keen interest in the issue, with Watson working now for the IPCC from a perch inside the White House. A second and stronger IPCC report in 1995 had made the case even more clearly: the world's delicate ecosystem was in danger of serious alteration from the buildup of carbon dioxide in the atmosphere, and the human "fingerprint" was clearly visible. "A pattern of climatic response to human activities is identifiable in the climatological record," the report said. In other words, human actions—in burning fossil fuels—were contributing to the problem. "There is no debate among any statured scientist of what is happening," summed up Harvard's James McCarthy, who was deeply involved in pulling together the second IPCC report, overseeing the work of the leading climate scientists from sixty nations.

With this kind of scientific clarity as a backdrop, the world met in Kyoto, Japan, in late 1997, continuing the process that had begun at the Rio Earth Summit five years earlier. Although debates went down to the final hours, a treaty to reduce greenhouse gas emissions was finally reached, with U.S. vice president Al Gore playing a crucial role in bridging differences between the developed and undeveloped worlds over where the burden of emission cutbacks would lie. No one was really satisfied with the final deal—it didn't go far enough, fast enough, or distribute the burden fairly enough—but just about everyone recognized what an enormous achievement it was nonetheless.

Inside the God pod, Exxon's top executives vowed to fight on, but it was hard not to be discouraged by the negative shift in momentum. The world seemed to be fighting back with surprising ferocity. This hadn't been in the original script! Sensing the strength of the public mood and the commitment of governments around the world, some of the biggest corporate players in the anti-Kyoto campaign started looking for an exit strategy. British Petroleum announced that, since it was now clear climate change was taken seriously by the broader community, the company was withdrawing from the anti-Kyoto Global Climate Coalition. BP was soon followed by Royal Dutch/Shell, Texaco, Ford, DaimlerChrysler, General Motors, Dow, DuPont and other former stalwarts. In the end, Exxon was left, clinging almost alone to the sinking ship, struggling to hold back the tide. All of a sudden it seemed that not even the endless resources of the fossil fuel industry were sufficient to stop the arrival of an idea whose time had plainly come. Like the Luddite attempt to stop automation and Big Tobacco's drive to discredit the case against smoking, the campaign against Kyoto seemed about to enter the history books.

———

Along with everyone else, Arthur G. Randol III had been surprised when the 2000 U.S. presidential election turned into a thirty-six-day nail-biter. But Randol was certainly delighted when the winner turned out to be George W. Bush, who had consistently sided with industry over environmentalists when he served as governor of Texas. As chief scientific adviser at Exxon, Randol knew that the Bush victory could only make the company's fight against Kyoto—and therefore his job—easier, but he didn't realize how much easier. Shortly after the Bush administration took office, Randol sent a fax to senior White House officials suggesting that the new administration remove Watson as head of the IPCC and replace him with someone more accommodating. It was a brazen request that he wouldn't even have bothered making to the Clinton administration, whose support for Watson had been solid.

The fact that Bush had actually won half a million fewer votes than his rival—not to mention whatever other mysteries remained locked in the sealed ballot boxes in Florida—might have made a different sort of man approach the top job in the world with a little humility, particularly on something as far-reaching in its implications as Kyoto. But Bush's feistiness made clear that the public's apparent preference for Al Gore's agenda would now become nothing more than an intriguing footnote in electoral history. In March 2001, just two months after taking office, the Bush administration announced its withdrawal from the Kyoto process, promising that the U.S. would instead tackle global warming in its own way. Washington's pull-out didn't kill Kyoto; indeed, the treaty came into effect in February 2005 after 141 countries signed on to it. But the withdrawal of the U.S., which produces 25 percent of the world's greenhouse gas emissions, was a devastating blow to the prospects for a meaningful international solution. "A fully effective climate treaty cannot work without the United

States," says Robert Watson. But Watson himself was soon out the door. In April 2002 the White House delivered on Randol's faxed request: it withdrew its support for Watson. Without the backing of his own government, support for Watson weakened and he ended up losing his position as IPCC chairman.

Meanwhile, in May 2001, the Cheney energy task force had released its report, which spelled out that America was hooked on fossil fuels and intended to stay that way. It called for greatly expanded subsidies for the fossil fuel industry. Speaking in Toronto that month, Cheney practically sneered at the concept of energy conservation. "Conservation may be a sign of personal virtue," he said, "but it is not a sufficient basis for a sound, comprehensive energy policy."

No longer taking on the world by itself, Exxon had found a friend. The most powerful government on earth had linked up with the richest corporation on earth—and the world no longer seemed invincible.

———

By the fall of 2002, the debate over Kyoto had flared up in Canada. Many Canadians were appalled by the attitude of the Bush government, and had agreed with Ralph Goodale, then Minister of Natural Resources, who responded to Dick Cheney by praising conservation as characteristic of "an advanced, intelligent society." Meanwhile, a feisty anti-Kyoto campaign was being orchestrated by the Alberta government and the Canadian oil and gas industry. Alberta premier Ralph Klein denounced Kyoto as "the goofiest, most devastating thing that was ever conceived," and the largely foreign-owned oil and gas industry kept insisting that investment in the Canadian oil patch would dry up if Ottawa were to sign the treaty.

As in the U.S., those resisting Kyoto in Canada had originally tried to create doubt about the strength of the science on global warming. But that had become progressively harder to do as the scientific case just kept getting stronger. (In June 2001, three months after Bush's rejection of Kyoto, the top scientific body in the U.S. came out in support of the central findings of the IPCC. A statement by the National Academy of Science noted: "The IPCC's conclusion that most of the observed warming of the last 50 years is likely to have been due to the increase in greenhouse gas concentrations accurately reflects the current thinking of the scientific community on this issue.") With the scientific case virtually unassailable, Kyoto opponents increasingly focused their efforts on claims that fighting global warming would devastate the economy. Canadians would be at a serious competitive disadvantage, they argued, if we committed ourselves to cutting back greenhouse gas emissions while our major trading partner refused to make the same commitment. There were charges that as many as 450,000 Canadian jobs could be lost—a figure that was widely bandied about and even posted on the Alberta government's website.

One might think that the possibility of such heavy job losses would be of particular concern to Canadian workers. But while the oil and gas industry and the Alberta government continued to warn of massive job losses, the Canadian labour movement was unconvinced by the dire predictions. In fact, virtually the entire organized labour movement in the country came out in favour of Kyoto. Even the Communications, Energy and Paperworkers Union, which represented 35,000 workers in the oil and gas sector, supported the global warming treaty, despite the fact that they would likely be the most affected by any reduction in oil patch investment. The 1,200 delegates to the union's national convention endorsed a pro-Kyoto stance, with only one dissenting vote. This was a curious situation, then: business opposed Kyoto because it said jobs would be lost, but workers—whose jobs were actually on the line—supported it.

Union official Fred Wilson explained that the union simply didn't believe the job loss predictions and suspected they were being trotted out as a scare tactic. Wilson noted that the energy workers' union was actually more worried about job losses due to other factors—like corporate downsizing and mergers, which had already eliminated 6,000 oil refinery jobs over the previous decade. (A joke doing the union rounds had it that, in order to reach the total of 450,000 lost jobs, the job toll would have to include all current energy workers and thousands of future energy workers, as well as thousands of dead energy workers, who'd lose their jobs retroactively.)

With the Canadian public and Parliament backing Kyoto, the Chrétien government ratified the treaty in December 2002, helping to keep it alive.

Interestingly, almost as soon as Ottawa made the announcement, the huff and bluster coming from the Canadian oil patch largely disappeared, giving credence to the argument that the campaign had mostly been a pressure tactic. In June 2003, *The Globe and Mail* ran a story in its business section headlined "Klein reverses Kyoto stand; now sees no peril to oil sands." The key threat to the province's oil sands development was apparently no longer Kyoto but—of all things—a labour shortage. "I don't know if [Kyoto] is much of a factor any more, because some compromise has been reached," Klein was quoted as saying. "But certainly the lack of workers—you know it's a phenomenal pace of growth—has slowed some of these projects down." Fears of 450,000 lost jobs had morphed into fears of there not being enough people to fill the existing jobs, and the "most devastating thing that was ever conceived" suddenly seemed rather harmless.

———

Having paid insufficient attention to car advertisements in the 1990s, I somehow didn't get the concept of an SUV. I didn't realize, for instance, that it was a symbol of a bold, adventurous, sporty kind of life, driven by people with a tendency to go off-road—just as, elsewhere in their lives, they have a tendency to break with the pack, to do things their way, to think outside the box. To show how far out of the loop I was, I didn't even know there was a difference between an SUV and a minivan. They both just seemed like awkward, bulky, oversized versions of a car—useful, no doubt, for those trips to Price Chopper when one comes home laden with several extra cases of Coke and a year's supply of toilet paper.

Of course, I was dead wrong. I've learned that there's a world of difference between a utilitarian minivan, which is designed for the Price Chopper trip as well as carting around children's soccer equipment, and an SUV, which is not only bold and adventurous but also glamorous, the car of choice these days among Hollywood stars and others with limitless resources. Still, one can appreciate the role advertising has played in making this sort of distinction clear to people, in establishing the SUV as the symbol of everything chic. The sheer brilliance of this advertising coup can perhaps best be measured by the extent of the image transformation the SUV has undergone since its first incarnation as a vehicle with few uses outside the funeral business. As Keith Bradsher has noted in his book *High and Mighty*, the forerunner of the SUV—the Chevy Suburban—dates back to the 1930s, and it managed to survive in the early decades largely because its body was the perfect height and width for the easy loading and unloading of coffins. This feature is retained today but omitted from SUV advertising.

So effective has the advertising campaign been that the public seems largely unaware that SUVs are generally difficult to handle, with poor agility and manoeuvrability—exactly the opposite of the sports car, which of course, used to be the sexy car of choice. While a sports car, with its low-slung body and road-gripping tires, can zip around corners at great speed, the rigid, high-set body of the SUV makes its way around corners with considerably more difficulty, which explains why ads typically show SUVs motionless at the top of a mountain or charging straight ahead (got to get straight home with all that toilet paper), rather than driving on the kind of winding, exotic cliffside roads typically seen in sports car ads.

Much of the appeal of today's SUV may have less to do with sportiness and glamour, and more to do with security in an age of fear. Huge and growing ever larger, the SUV offers its riders a massive, wraparound steel exterior with the feel of a tank—a mobile version of a gated community. Bradsher reports that this is deliberate, that the tough, even menacing-looking appearance of many SUVs is intentionally and consciously designed by automakers for an era when civility on the roads has been replaced with unabashed hostility, a kind of me-first aggressiveness. In the age of everything from road rage to anthrax to SARS, you can't have too much steel between you and the rest of what's driving around out there.

And it's true: SUVs *are* a threat to others on the road—a fact that was appreciated as early as thirty years ago by researchers trying to draw attention to the dangers of designing vehicles with the sort of stiff, high front ends that are the hallmark of the SUV. In an accident, an SUV is two and a half times more likely than a mid-size car to kill the occupants of the other vehicle, due to the fact that SUVs are heavier (by a thousand pounds on average), stiffer and taller. These characteristics effectively turn SUVs into "battering rams in collisions with other vehicles," notes a recent report by the Union of Concerned Scientists. When an SUV hits a car from behind, it is more likely to ride up over the car's bumper, leaving the car (and its occupants) essentially defenseless. When an SUV hits a car from the side, it can ride up over the car's door frame and right into the passenger compartment. (Well, hello there!) This mismatch of sizes has been dubbed "vehicle incompatibility," but another possible name would be "vehicle homicide." And one-on-one against an unarmed pedestrian, an SUV is more lethal still, hitting the pedestrian higher up on the body, closer to vital organs, than a regular car does. Yet this greater height, stiffness and body weight seems to be something of a selling point, rather than a signal that perhaps something is seriously wrong with these oversized killing machines.

Even if the thought of killing others isn't a deterrent, one would think the thought of killing oneself and one's loved ones would count for something. But apparently not. SUV sales have skyrocketed despite the fact that they are also a danger to their own passengers because, with their considerable height, they have a tendency to roll over. More than 51,500 occupants of SUVs (and light trucks) died in rollovers from 1991 to 2001. The overall fatality rate for SUVs was 8 percent worse than for cars in 2000.

But the more far-reaching problem with SUVs—at least in terms of the survival of the planet—is the devastating amount of greenhouse gases they spew out into the air. An SUV produces roughly 40 percent more greenhouse gas emissions than a regular car does, and with SUV sales soaring—sales have increased seventeen-fold in the past two decades—their emissions have become a significant part of the problem of global warming. At a time when the dangers of global warming are blatantly evident and of serious concern to people all over the world, the breezy marketing of SUVs in North America makes a mockery of any claim that we are addressing the problem. While common sense would call for a special effort to move away from these over-emitting vehicles, exactly the opposite has been happening. Both the U.S. and Canadian governments have contributed to the SUV problem by offering SUVs regulatory controls far looser than those applied to regular cars. (The regulations were established in the U.S., but

Canada has effectively adopted matching standards, which the automakers have agreed to deliver on a voluntary basis.) The extraordinary growth in SUV sales over the last two decades, then, can be attributed as much to government favouritism as to the massive advertising campaign that has left prospective SUV buyers thinking of off-road adventure rather than the ease of moving coffins.

The story of the SUV is in many ways a microcosm of the human folly that has led us to the brink of a climate change disaster. Perhaps it seems unfair to pick on the SUV. After all, it isn't, by any means, the only source of greenhouse gas emissions. It is, however, one of the fast-growing sources. SUVs now account for an astonishing 24 percent of all new cars sold, up from just 2 percent in 1980. (Overall, the transportation sector accounts for 26 percent of U.S. greenhouse gas emissions. Along with coal-fired power plants, transportation is one of the key sources of the global warming problem.)

I'm picking on the SUV partly because it somehow serves as a metaphor for the absurdity of the situation we find ourselves in, if for no other reason than that these enormous, awkward vehicles seem so . . . well . . . unnecessary. There's another aspect to this story that makes it emblematic of the saga of global warming: how easily the problem could be corrected if there were any serious political will. The technology exists to make enormous strides in cutting back the greenhouse gas emissions currently spewing out of SUVs (and other vehicles, but particularly SUVs) on highways across North America. I'm not talking about exotic space-age cars that run on hydrogen in some dream scenario (that's likely a couple of decades or so down the road), but about technology that already exists—and currently sits on shelves—in the offices of our big automakers. This, then, is the story of how Luddites in the auto sector, fearful of risking their dominant market position, have declined to take us to where any sane person can see we must go, hiding behind claims of technological "can't do," hoping the public won't realize that what we have here is, in fact, a tale of technological "won't do."

————

The SUV tale actually begins with a stunning success story. When events in the Middle East led to dramatic oil price hikes and a brief oil embargo in the mid-1970s, the Western world was confronted with an energy crunch, and in some ways at least, it responded sensibly. Western governments, including those of the U.S. and Canada, dealt with the oil shortage in a logical, effective way: they got us to cut back our overall consumption of oil. National efforts were launched to encourage people to conserve energy. Part of this involved simply making people aware of things they could do, from turning off the lights to weatherstripping their homes. Part involved mandatory changes, like new regulations requiring appliance manufacturers to use more electricity-efficient technology (which didn't even turn out to be more expensive). The result of this drive towards energy conservation was that overall energy usage quickly dropped by about 30 percent, consumers saved billions of dollars a year in energy costs and no one was really left worse off—unless one considers having more efficient refrigerators a setback.

One of the most effective new measures involved cars. Given the significant amount of oil they consumed, cars had to be part of any meaningful energy conservation program. So, sensibly, in 1975 the U.S. introduced what became known as CAFE (Corporate Average Fuel Economy) standards, under which automakers were required to make their cars more fuel efficient. There was resistance to the CAFE standards from the automakers, and dire predictions of the impact the new rules would have on the industry, but Congress persevered, backed by strong support from the public, as just about everyone recognized the seriousness of the country's energy predicament and considered action of this sort necessary. The surprising thing was that no one had really bothered to pay attention to fuel efficiency before. The existing fleet of North American cars was hugely energy inefficient, with cars typically getting no more than 13 miles to the gallon. With a strong sense of national purpose, CAFE standards were enacted requiring automakers to produce fleets of cars averaging 27 miles to the gallon—twice their previous level—by 1985.

In response, automakers quickly achieved significant improvements in energy efficiency. Chrysler was a particularly notable success story. By 1985, its fleet of cars was averaging 27.5 miles per gallon, as required under the CAFE rules. The company reported that it had invested $5 billion in meeting the standards. It also reported that *it had never been more profitable*! Clearly, the goal of greater fuel efficiency was achievable, and corporate balance sheets didn't even have to suffer. The next step—setting higher CAFE standards—should have been easier, now that Chrysler had shown the way. But since the energy crisis had subsided, the political will had largely evaporated too. General Motors and Ford had made significant strides towards fuel efficiency but were falling short of the existing standards, for which they could have been heavily fined. But they were not.

They argued that the standards were too high, and succeeded in convincing Congress to roll back the standards, to 26 miles per gallon. Among those most angered by the rollback was, understandably, Chrysler.

Chrysler's critique of the position of its fellow automakers offers a rare insight into how easily such standards can be met. "Our compliance with the [CAFE] standard is proof that the 27.50-mpg standard is technologically feasible and that other manufacturers could have met the law as well," Chrysler president Harold R. Sperlich told a U.S. House subcommittee in September 1985.

Chrysler chairman Lee Iacocca went further in an interview with the *Chicago Tribune*, ridiculing claims that if the other automakers had been forced to pay fines for not meeting the CAFE standards, they would have had to shut down plants and lay people off: "Would GM shut a plant because, instead of making $5,000 profit on a car, they had to pay a CAFE fine and only make $4,500? That's mad; that's crazy." (Another option would have been for the two wayward automakers to meet the standards.) But in rolling back the standards as requested, Congress set a precedent: it would give in on this issue under pressure. Iacocca, with considerable foresight, noted at the time: "We are about to put up a tombstone—'Here lies America's energy policy.'"

In fact, there's been no real progress since then, despite the crucial realization in the intervening years that there is a serious threat of climate change and that cars are a major source of this problem. In 1990—even as the world was planning to come together at the Earth Summit to address the global warming issue—the U.S. Senate responded to intense auto industry pressure and rejected a bill that would have gradually raised CAFE standards by another 40 percent. Six years later, with the world increasingly aroused by the dangers posed by global warming, the U.S. House of Representatives showed its determination to take no further action when it inserted a rider in an appropriations bill freezing CAFE standards at their existing levels. To prevent the Clinton administration from pursuing the issue, the rider also banned the federal Department of Transportation from studying the feasibility of higher standards, or even considering the need for them! Tougher CAFE rules weren't just off the table, Congress wanted to make sure that nobody would even think about putting them back on. And in 2003, after the Bush administration had turned its back on the Kyoto process, Congress once again rebuffed moves to tighten CAFE standards, adding new hurdles to future attempts to move in that direction.

Canada has been similarly negligent. Since the car market
in North America is integrated, with the big automakers producing cars on both sides of the border, Ottawa has accepted
the U.S. CAFE standards and allowed the big automakers to vol-
untarily comply with them here. That initially worked well,
since it meant cars sold in Canada had to meet improved fuel
efficiency standards, just as cars sold in the U.S. did. But after
the U.S. abandoned the quest for higher standards in 1985,
Canada went along with this sorry development too. Ottawa
could, of course, have imposed its own tougher standards.
Indeed, in 1981, the government of Pierre Trudeau passed the
Motor Vehicle Fuel Consumption Standards Act, which gave
Ottawa the power to do just that. Under pressure from the
automakers, however, the Trudeau government never actually
put the law into effect.

While Canada had a history of inaction on the issue, it appeared
for a while that that might change after Ottawa signed the Kyoto
accord in late 2002. With growing pressure from the public and
environmental groups to do something about global warming,
Ottawa indicated that it was considering toughening up its fuel efficiency standards by 25 percent. The auto industry reacted
negatively, complaining that the small Canadian market wouldn't
justify making the necessary changes just to accommodate tougher
Canadian laws. But, in fact, the proposed tougher Canadian laws
being proposed were modelled on tougher laws already introduced
by the state of California. Together Canada and California make up
a market of 62 million people—significant enough for the car companies to be forced to accommodate it. The companies had gone to
court to try to quash the California regulations, but having Canada
join forces with California would certainly weaken the companies'
manoeuvring room.

(This was part of a larger debate about how Canada was to
meet the greenhouse-gas-emission reductions set out in the

Kyoto accord, which require Canada to reduce emissions by 6 percent below 1990 levels by 2012. Meeting the terms of Kyoto is a significant task, especially since Canada's emissions have grown by more than 16 percent since 1990. Even so, there have been some encouraging developments: emissions from households, commercial buildings and most industries have grown very little in recent years. The problem is that certain sectors—notably the fossil fuel industry and the transportation sector—have grown substantially. It's hard to see how any plan that fails to deal with these sectors can provide a meaningful solution. But Ottawa's Kyoto plan, unveiled in April 2005, placed the burden of reducing greenhouse gases heavily on individual households; the major polluters were largely let off the hook.)

In the end, Ottawa settled for a "voluntary" deal with the auto industry, leaving it up to the discretion of car makers to make improvements in fuel efficiency. Even with the pressure of meeting Kyoto targets, with strong support from the Canadian public and the potential for an alliance with California, the Liberal government of Paul Martin had backed off from taking a tough stand with the auto industry.

The failure to make any advances in the fuel efficiency of cars for almost two decades would be understandable—if the technology to make further improvements didn't exist. But this isn't the case. Notes David Friedman, a mechanical engineer and specialist in transportation technology who works for the Union of Concerned Scientists in Washington: "This is technology that the automakers have developed themselves." And they're using it—but for other things.

Greater fuel efficiency simply means that it takes less energy to move a vehicle down the road. Engines can be made more efficient, for instance, by designing them to reduce internal friction and to ease the entry of air so that the fuel can burn more completely. With these sorts of improvements, vehicles can be propelled down the road at the same speed while consuming less fuel—hence, fuel economy. But greater engine efficiency doesn't have to be put towards fuel economy; it can be put instead towards other advances, such as propelling a bigger vehicle down the road at the same speed (or faster). And this is essentially what has happened. In the absence of tougher CAFE standards, automakers haven't had to achieve greater fuel economy. So, *while they've continued to make engines more efficient*, they have put these efficiency gains towards other things— primarily towards achieving ever greater propulsion for ever larger vehicles. As a result, they have been able to build larger, heavier vehicles that have the same propulsion as smaller, lighter vehicles. It's hard to see this as anything other than the foolish squandering of significant technological gains. "If we keep on the path we're on," notes Friedman, "we'll be seeing 18-wheelers that accelerate like racing cars."

————

The failure to take tougher action against fuel-inefficient cars is a sorry tale, but it's a sideshow to the really tawdry part of the story—the special case of the SUV. SUVs are exempt from the CAFE standards that apply to cars; they have their own, much less rigorous CAFE standards. They are permitted to release about 40 percent more greenhouse gases per vehicle into the air. This has created a lethal dynamic. With lower fuel efficiency and skyrocketing sales, SUVs have quickly become a key—and seemingly needless—part of the global warming problem.

The original decision to exempt SUVs from CAFE standards for cars was actually fairly innocuous. Back in the mid-1970s, SUVs barely existed; they accounted for only 2 percent of the vehicle market. In fact, the decision to create a separate category for the purposes of CAFE standards had little to do with SUVs. It was really about creating a lower standard for light trucks, that is, pickup trucks and commercial vans. There was some logic to this.

Trucks and vans often have to cart around heavy loads, and therefore arguably need to be bigger and heavier. This means that their engines require more energy (and therefore more fuel) to propel them down the road. So, it was argued, it would be unfair to require these work vehicles to operate on the same amount of fuel as cars, which carry around lighter human loads.

The problem arose with the willingness of Congress to be talked into classifying certain large passenger vehicles (later called SUVs)—like the Ford Bronco, the Chevrolet Blazer and the Jeep Wagoneer—as light trucks rather than cars. These vehicles were essentially oversized cars, not work vehicles. They were hauling around relatively light loads of passengers, not concrete or heavy equipment. Therefore, there was really no reason to give them an exemption on fuel efficiency, except to make things easier for the automakers. (The CAFE standards are averaged out over the entire fleet of cars produced by an automaker, so if these heavier models were included in an automaker's car fleet, the rest of the fleet would have to be that much more fuel-efficient to compensate.)

By allowing SUVs to have lower fuel efficiency, Congress was allowing them to remain big and heavy, even though they didn't really need to be, while cars were generally becoming smaller and lighter, in order to meet higher fuel-efficiency standards. Thus was created a size gap that would become a haunting safety problem. Indeed, there was also an incentive for SUVs to become even more massively heavy. If they weighed in above a certain gross vehicle weight, they could qualify as *larger* light trucks (presumed to be a work vehicle for hauling things)—and then escape fuel efficiency regulation entirely! So, if they could just be bulked up to six thousand pounds (a stunning size), they could burn as much gas as they wanted. Nobody could stop them from being hopelessly fuel-inefficient, even getting as little as one mile to the gallon. Imagine the freedom!

Now, one question that might pop up here is: why do people want such gigantic vehicles? A bigger size might make sense for large families, but families are typically getting smaller, and the appeal of SUVs is by no means confined to the family market. No doubt part of the quest for bigness can be connected with the notion that bigger is better—an idea that certainly thrives in our culture. Why wouldn't one want more of something if one could have it, in the same way everyone apparently wants a bigger house and a bigger serving of popcorn at the movies? (Popcorn servings have typically doubled in size over the last couple of decades.) Driving around in a great big 6,000-pound SUV, towering over others, presumably makes the owner look awfully impressive, awfully in command of the road and, by extension of his or her life—not to mention the lives of those driving around below, who to some degree continue to exist at the whim of the SUV driver.

Allowing SUVs to grow ever bigger has created a significant size gap between them and regular cars, which, as we've seen, is dangerous to people in regular cars. And this would presumably be further reason (in addition to fuel efficiency) for cancelling the more lenient CAFE regulations for SUVs. But the U.S. Congress has moved in exactly the opposite direction. Rather than eliminate special treatment for SUVs, it has created a number of additional special exemptions favouring these dangerous, earth-destroying vehicles. The original reason for this extreme bias in favour of the SUV seems to date back, strangely enough, to a trade dispute some forty years ago. It's a bit of a detour (it involves poultry!), but worth exploring briefly if only to illustrate the arbitrariness of what we're dealing with here.

In the early 1960s, Washington was angered when European countries imposed high import taxes on frozen chickens from the United States. Washington took its complaint to an international trade panel in Geneva, which ruled in favour of the U.S.

and permitted Washington to retaliate with import taxes (equivalent in value to the losses American farmers suffered because of the European chicken tax). Accordingly, Washington slapped steep import taxes on a few carefully chosen items, including high-priced brandy (to punish the French) and light trucks (to punish the West Germans, particularly the West German carmaker Volkswagen, which was selling vans and pickup trucks in the U.S. market). Under the rules of international trade, these import taxes had to be applied to the same goods from other countries as well. Over time, the original dispute got resolved and the tax on high-priced brandy was removed. But the 25 percent import tax on light trucks has remained in place, and it's not just Volkswagen that has suffered.

The big impact, ironically, has fallen on the Japanese. In the decades that followed the implementation of the tax, Japanese automakers became highly competitive, developing top-notch cars that quickly captured a significant portion of the U.S. market. The fear this generated in U.S. automakers was captured graphically in a 1971 taped conversation in which Lee Iacocca told President Richard Nixon: "We are on a downhill slide the likes of which we have never seen in our business. And the Japs are in the wings ready to eat us alive." One small portion of the U.S. market where the Japanese didn't make any serious inroads was light trucks, because of the 25 percent import tax they faced. So, while the Japanese grabbed a significant share of the Big Three automakers' North American market for cars in the 1970s and 1980s, they barely gained a toehold in the market for light trucks, which belonged—lock, stock and barrel—to the big Detroit automakers.

It is not surprising, then, that Detroit concentrated its development and promotional efforts in the light truck market—a market where it was effectively sheltered from foreign competition. Protected by a high tariff wall keeping out foreign light trucks, this was a sandbox where only U.S. manufacturers could play! Unlike the other sandbox, where tough foreign competitors were increasingly taking up space, here Detroit automakers had things all to themselves. The light-truck market became their security blanket, their sheltered workshop. The technologically sophisticated Japanese might whoop them in the car market, but here, in the protected confines of the light-truck market they could grow and prosper, safe from the rigours of the global economy.

All this would have been pretty inconsequential if the light-truck market had really just involved light trucks, vehicles of interest mostly to farmers and contractors. But with the decision to classify SUVs as light trucks, the potential for growth of this market became huge. If enough consumers could be encouraged to switch from cars to SUVs, the light-truck market would swell to a significant proportion of the overall vehicle market, helping the Big Three to maintain their dominant position.

But could this be done? Could consumers be persuaded to see the SUV not as some clunky truck-like thing, but as a chic, cool, even sexy version of the car? The Big Three began sinking enormous resources into promoting such a makeover, turning the SUV into one of the most heavily advertised products in North America. It was pictured on mountaintops, on sprawling estates, on highways towering over cars. Its bloated body was presented as macho, as threatening to smaller, unprotected vehicles trying to claim a space on the same road. In one ad, in which a big SUV and a little car are on the same stretch of roadway, the cutline reads simply "YIELD." Arnold Schwarzenegger, in his movie star days, was hired to drive General Motors' civilian SUV version of a Humvee military vehicle into Times Square. More than $9 billion was spent on SUV advertising from 1990 to 2001, with $1.51 billion spent in 2000 alone (out of an entire U.S. advertising market in all sectors of about $90 billion). All this was directed towards encouraging the notion that SUVs are ultra-cool, not large, hard-to-handle vehicles suitable for the transport of coffins and bargain-sized toilet paper packages. By promoting the improbable notion of the SUV as a chic car—and with enough money, apparently, even a makeover this far-fetched is possible—the Detroit automakers were able to create an aura around SUVs, making them the most desirable of vehicles, generating a hunger for them at all income levels.

The big automakers had little trouble getting government onside. In asking for special treatment for SUVs, the Big Three were essentially asking Congress to give a legup to domestic manufacturers (and workers), who were finding it hard to hold on to their traditional dominance of the North American car market in the face of intense competition from the Japanese. Despite all the high-minded trumpeting about the virtues of free trade and open competition in public speeches, legislators were only too happy to use a sneaky route to give an advantage to the U.S. auto industry. Thus, whenever new regulations were introduced, they were first applied to the car market, where real international competition existed, and only later, and in a watered-down form, to the light-truck market, where Detroit dominated. (By the late 1980s the Japanese had become so competitive that they began shipping SUVs into the American market and managing to compete—even after paying the 25 percent import tax.)

Still, the light-truck market remained largely American turf, and so light trucks were awarded many exemptions—starting back in 1973, when the U.S. Environmental Protection Agency (EPA) allowed American Motors' Jeep to be classified as a light truck and thereby escape regulation under the air pollution rules of the Clean Air Act. Two years later, as we've seen, Congress spared SUVs the tough new CAFE fuel economy regulations for cars, making them subject instead to the looser regulations for light trucks (or the non-existent regulations for "larger" light trucks). In 1978, Congress enacted a tough new measure to zero in on vehicles that were still burning large amounts of gasoline. Known as a tax on "gas guzzlers," the measure (still in effect) adds more than $7,500 to the price of some high-powered sports cars. In fact, these sports cars guzzle quite a bit less gas than a standard SUV, but, astonishingly, SUVs—because of their status as light trucks, are exempt from the gas guzzlers tax.

SUVs were favoured once again in 1984, when Congress tightened tax rules under which self-employed professionals and business people were permitted to deduct costs related to vehicles they used for work. Under the new rules, deductions for car purchases were limited to $17,500 and were to be deducted over a longer time period. But none of this applied to the SUV. Its full price (up to $38,200) could be deducted, with almost half deductible in the first year. So the tax code provided an enormous incentive for the self-employed not only to switch from cars to SUVs, but to switch to the most expensive SUV models (thereby maximizing their tax deduction), which also tended to be the biggest gas guzzlers. And if this wasn't incentive enough, in 1990, when Congress imposed a 10 percent luxury tax on cars with price tags above $30,000, SUVs were again exempted.

When the luxury tax was introduced, cars made up 95 percent of the luxury market. Six years later, with cars burdened under the new tax, a whole new breed of huge, luxurious, pricey SUVs—exempt from the luxury tax and providing enhanced deductibility from regular taxes—had triumphantly moved onto the scene, claiming fully half of the luxury market.

———

In the reception area, there is a large display of material promoting energy conservation, full of sensible tips about turning off the heat in unused rooms, properly insulating windows, etc. It's easy to forget this is the Washington office of the group with possibly the most to lose if people actually followed such advice.

This is the office of the American Petroleum Institute (API), the lobby group for the petroleum industry. Its member companies

obviously make money from our obsessive attachment to petroleum, so it would seem more logical for them to have promotional
displays urging us to crank up the heat, avoid the nuisance of
weatherstripping, etc. But then, this is an industry that has taken a
fair bit of criticism over the years, and has learned to adopt a public posture of social responsibility on the energy question. So
energy conservation is duly celebrated in the API reception area,
even while the API does its best to undermine any meaningful
approach to energy conservation in the legislative arena, where it
might actually make a difference.

The three API executives I met with would fiercely contest this
description of their activities, preferring to present themselves as
merely trying to bring some cool-headed thinking and reflection
to the debate on global warming. They are careful to sound reasonable, to appear to be simply asking for more certainty in the
science, questioning whether there hasn't been too much oversim-
plification of a complex subject. "I don't see science used
productively," says Robert Greco, the API's director of global climate programs. "Science advances when someone hypothesizes
and is proven wrong and then revises their hypothesis. But there
didn't seem to be room for that in the climate science debate."

Yet this sort of approach—in which a scientific hypothesis is
advanced, critiqued and revised one small step at a time—
appears to be exactly the one adopted by the IPCC, operating
under the auspices of the UN and involving thousands of scientists around the world. The API executives complain that the
IPCC has ended up taking a point of view. (What else should it
do, when the evidence of man-made global warming is apparently so overwhelming?) They also complain that the debate has
become polarized. This is true, although the "polarization"
seems to consist of virtually the entire scientific community in
one corner, and a handful of skeptics and the petroleum industry in the other. Greco attempts to portray the API as a voice of

FROM COFFINS TO WORLD DESTRUCTION   179

moderation, trying to bridge the gap between two extremes. "The truth is probably somewhere in the middle," he says. But is that necessarily so? Is the truth always inevitably somewhere in the middle? Forced to choose between the theory that the earth is flat or one that says it is round, should we assume the truth lies somewhere in the middle and conclude that the earth is cylindrical?

Given the uphill battle it faces, the API's approach has been reduced to running interference, to casting doubt in the public's mind, to creating uncertainty where it needn't exist—ultimately, to throwing sand in the wheels of the enormously complex project of coming up with a global solution to climate change. Part of the strategy involves moving the issue away from the strict confines of the scientific debate, where things are carefully measured and quantifiable and where it's therefore more difficult to twist the truth. Instead, the debate has been shifted into the far more nebulous terrain of economics, where assertions with little evidentiary support often pass for serious commentary (for example: "The private sector is always more efficient than the public sector," "Giving rich people more money will benefit the whole economy," etc.). The API has made its counterattack mostly in economic terms, suggesting that we can't really tackle global warming without inflicting enormous economic hardship on ourselves. Russell Jones, the API's research manager (and an economist), says that serious reductions in greenhouse gas emissions can't be achieved without crippling the economy. "We can't do that with today's technology and keep people going to work." The limitations of today's technology seem overwhelming to the folks at the API.

Across town at the Union of Concerned Scientists (UCS), where there is considerable expertise in today's technology, no such gloom prevails. In fact, there is boundless confidence about what can be achieved—*using nothing more than current technology.*

Engineers working for this non-profit group of scientists have already designed their own SUV—the UCS Guardian, they call it—which is the same size and has the same power as a typical SUV but is 30 percent more fuel efficient. As the group notes, "All the technologies and design techniques employed in the Guardian are available in mass-produced vehicles in the United States today." The Guardian's engine, for instance, incorporates a 225-horsepower, 3.1-litre, dual overhead cam V6 engine with four valves per cylinder and variable valve technology, along with low-friction design and engine oil, and its body is more aerodynamic in design and uses higher-strength steel, which allows it to weigh less. "This isn't rocket science," adds David Friedman, one of the engineers who worked on the SUV. In fact, it's just a patchwork of different features that are already being used in cars driving around on today's roads but are not being used in SUVs.

These improvements would add about $600 to the price of an SUV—a tiny additional cost for purchasers who don't seem particularly fazed by high prices and who are often deducting the full cost of the vehicle anyway. Furthermore, once the gas savings are thrown into the equation, the fuel-efficient SUV ends up being a money-saver over time. (The scientists couldn't resist adding a few safety improvements to their SUV as well, such as a stronger roof, for an extra $50, which would protect the heads of occupants during a rollover.) In other words, it's possible to have all the design features of an SUV (big, bulky, imposing) without guzzling 40 percent more gas than a car does. And none of this would involve people staying home from work, or even altering their driving patterns or habits.

Of course, this isn't by any means the solution to climate change; it would simply be a small step in the right direction. Ultimately, among other things, we have to move away from the internal combustion engine altogether, adopt the advanced fuel technologies currently under development and learn to rely far more on mass transit. But while we're on the subject of the regrettable state of our current vehicles, let's just acknowledge that things could be a whole lot better than they are, without any noticeable suffering on the part of the general public.

Inevitably, the sheer simplicity of this solution seems to make people skeptical, even suspicious. Comments Friedman: "People say, 'If it's so simple, somebody would be doing it.'" Back at the API, the executives say almost exactly that: "If it's so cost-effective, why don't the automakers put it in?"

Then one of the API executives offers an answer: "Because they're big, nasty people!"

At this, the API executives explode with laughter. This is clearly the kind of payoff moment they've been waiting for, and they relish every second of it. Surrounded by their echoing hilarity, I feel somewhat silly, at least on the defensive. My line of questioning had clearly been aimed at suggesting there were solutions available that weren't being used. With the "big, nasty people" line, the API executives are firing back, taking the offensive, derisively suggesting that my point is, essentially, a conspiracy theory: only the presence of evil people in the auto companies could possibly explain why these obvious solutions were being resisted.

I'm not quite sure how to respond.

One thought crossing my mind is that it's possible that those running the auto companies *are* big, nasty people. Such people do exist. On the other hand, I doubt those running the big car companies are big and nasty. Rather, I suspect that, as individuals, they are personable and pleasant, and would be most hospitable if you somehow found yourself at their dinner table. But they're part of an industry that has adopted an obstructionist position on meaningful solutions to global warming—largely, I suspect, because they think these solutions would interfere with the profitability of their companies and they consider it their responsibility to ensure that profitability. So, while an individual auto executive may think in the abstract that it's his responsibility to address global warming *as a citizen*, he knows it's his responsibility to protect the company's profit *as a corporate manager*. On the citizen front, he is simply one among billions in the world who share this responsibility. On the corporate manager front, it boils down to him and a few others. And his neck, he figures, would be very much on the line if company profits fell, whereas his neck would be only one of billions potentially on the line if the earth's temperature got too hot (which would probably not be in his lifetime anyway). And although the negative consequences of global warming for the earth and its inhabitants are undoubtedly greater in the overall scheme of things, the negative consequences of his company losing money loom larger in his particular life.

His inclination to focus more on the corporate profitability problem is reinforced by all sorts of comforting arguments: the science isn't clear; the impact on the economy would be devastating; consumers aren't all that interested in fuel efficiency. His job, after all, is to give consumers what they want (that's what capitalism and free markets are all about). Furthermore, once the notion develops out there that auto or oil executives are big, nasty people, there's an understandable tendency for them to become defensive, to circle the wagons, to resist being pushed around, to hold their heads high in the face of criticism and personal slights. Hence the joviality among the API executives when they can strike back against a do-good journalist arguing about saving the earth's crust from more warming (easy for her to do—she isn't dependent for her personal advancement on those who own stock in oil companies).

So they resist. They argue against regulating fuel efficiency, insisting that more can be accomplished through voluntary co-operation than through mandatory requirements. It's interesting to note that the auto industry also resisted compulsory turn signals, seat belts and air bags, and that it lobbied successfully to have SUVs exempted from regulations requiring headrests to prevent whiplash and steel door beams to minimize injuries from side collisions. (Another victory, presumably, for SUVs.) As we learned with turn signals, seat belts, air bags and other safety features, these only really became a regular part of the auto fleet when manufacturers were required to install them, and then the whole thing turned out to be easily done, with none of the dire predictions of hardship coming true. In 1966, Henry Ford II argued that proposed federal safety requirements, including compulsory seat belts, would force his company to "close down." As we know, it's still around and doing fine.

This is pretty much the pattern with all new environmental regulations, according to Elizabeth May, who has observed the process many times. She argues that there are even recognizable stages corporations go through in resisting new regulations: first they deny the science, then they deny they're the ones causing the problem, then they insist the economic damages in tackling it would be catastrophic. "Then once they see that the government is serious, there is acceptance—and even increased profitability," she sighs. "It's all very predictable."

One thing is clear: the voluntary approach won't work with fuel efficiency. The automakers are just too focused on making ever bigger, taller, mightier vehicles, catering to a market that appears to crave size and apparent security (if not actual safety). And the bigger some SUVs get, the more others are bulked up too, to keep pace—leaving people driving around in mere cars ever more vulnerable. The dynamic leads to bigger and bigger. Friedman likens it to the Cold War concept of Mutually Assured Destruction (MAD): only when each side armed itself equally to the teeth could the risk of nuclear war be minimized, because neither side would then be willing to bring about its own certain destruction. With the MAD mentality reigning on North American highways, arming oneself with an ever bigger vehicle almost seems like the only responsible way to protect one's family from that monstrous and menacing SUV visible in the rear-view mirror. And what about those who can't afford a 6,000-pound vehicle to properly defend themselves on the highway? Well, it's a free world, and there's always the option of staying home.

The logical way to break this pattern would be for governments to implement tougher CAFE standards, following up on the success achieved on this front in the 1970s. With tougher fuel economy requirements—applied to both cars and "light trucks"—automakers would have to focus on applying their technological advances to making more fuel-efficient engines, rather than making ever heavier vehicles with peppier performance. And the impact on fuel consumption, according to the Union of Concerned Scientists, would be dramatic. The scientists compared this mandatory approach with a voluntary approach, based on the commitments that the automakers have publicly pledged to achieve. Even assuming that the automakers live up to these commitments, the oil savings would be minimal. By the year 2020, the U.S. auto fleet would be consuming just under 12 million barrels of oil a day, as opposed to just *over* 12 million barrels a day (the projected use if current patterns continue)—a saving of just 4 percent. But under a realistic scenario of stronger CAFE rules for both cars and light trucks, oil consumption would drop to about 7.5 million barrels a day—a saving of about 40 percent, or ten times the savings achieved under the voluntary plan. "The most important action we can take to reduce our dependence on oil is to reinvest in the CAFE standards," the UCS argues.

Of course, we'd have to forgo the dream of having 18-wheelers that accelerate like race cars.

A farm boy from Dearborn, Michigan, Henry Ford was a self-taught mechanical genius with a feel for building cars. But his rise to prominence and dominance in the early days of the auto industry, at the beginning of the twentieth century, when there were dozens of small start-up car-building companies, revolved around his celebrated invention of the assembly line. While other car companies were raising the price of their cars a bit each year as the technology improved, Ford was efficient enough—and smart enough—to lower his each year. A Model T Ford cost $950 in 1909, a substantial sum of money. By 1925, when the Model T was a much better car, it had dropped in price to only $240. Such savings didn't go unnoticed by consumers. By 1918, nearly half of all the cars in the world were Model Ts. And the relatively high wages earned by Ford assembly-line workers meant there was a growing market for the thousands of cars rolling out of the Ford factory.

If this concept of affordable cars for ordinary people captured the spirit of an earlier age, the proliferation of increasingly pricey SUVs for an ever more upscale market seems to capture the spirit of ours. But there's another way in which the rise of the SUV could be seen as a repudiation of the legacy of Henry Ford. One of the reasons that SUVs have been such a source of joy to Detroit automakers—aside from the special exemptions for them handed out by legislators—is the fact that they are much cheaper than cars to produce, *largely because they have an inferior construction*. So, while Ford produced ever more technologically advanced vehicles for ever lower prices, today's SUVs are more technologically primitive vehicles—for ever higher prices. Now there's a bargain not to be missed.

Since the late 1970s, cars have been constructed in a more technologically advanced way, with the underbody and the side panels being all one unit. Among other benefits, this provides "crumple zones" that absorb the impact of a crash, offering greater safety for the occupants as well as others on the road. SUVs, on the other hand, are constructed using a technology—about one hundred years old—in which a body frame is imposed on a separate underbody. While this construction provides extra strength, which is useful for towing heavy loads, it has the disadvantage of being heavy (and therefore poor in terms of fuel economy) and stiff (and therefore less inclined to crumple well, making everyone inside and outside the vehicle less safe).

But this more primitive design is cheaper to produce, and it has allowed automakers to make far bigger profits on SUVs than on any other vehicles. Churning out SUVs on a round-the-clock basis, Ford's Michigan Truck Plant had, by the late 1990s, become the single most profitable factory in the world. Although the company had more than fifty assembly-line factories around the globe, the SUVs streaming out of the Michigan plant accounted for a whopping one-third of all profits earned by Ford in 1998. It's easy to see why. Each Lincoln Navigator, selling for $45,000, delivered a net profit of about $15,000. That trend has continued. By 2003, George Peterson, president of AutoPacific Inc., a California-based auto industry consultant, estimated that a $50,000 SUV was delivering a profit of $20,000 to its maker—roughly ten times the profit on that sexless, utilitarian vehicle, the minivan. "Fortunes were waiting to be made with each new SUV model that would be a little bigger and a little more luxurious," notes Keith Bradsher.

The technological backwardness of the SUV may also help explain why Japanese automakers, although now producing SUVs in U.S.-based plants, have been unable to become dominant in this corner of the market. Perhaps the knack of the Japanese for developing increasingly sophisticated technology is working against them here. After all, the SUV isn't about more sophisticated technology. It's about exactly the opposite. It's about sticking with more primitive technology, about failing to adapt the technological breakthroughs of our times to the environmental crisis of our times. It's the Luddite dream car—clinging to yesterday's technology in the midst of today's looming disaster. Could it be that there's something so foolishly oversized, graceless, clunky—and flagrantly indifferent to the public interest—about the SUV that only a North American company could fully appreciate how to design one?

———

The spectacular rise of the SUV could be seen as a series of fluky developments. If it hadn't been for the 1960s trade dispute over chickens and the resulting import tax on light trucks, the protected market for light trucks—free of ever inventive Japanese competitors—might never have come into existence. Without a protected market, U.S. automakers would have been less inclined to develop and promote huge vehicles under the guise of light trucks, and U.S. lawmakers would have been less likely to grant these oversized vehicles favourable treatment. So, in a sense, the whole thing could be chalked up to a mistake that got out of control. Not to big, nasty people.

Whether mistake or not, however, by 2003 there was no mistaking that the situation *was* out of control. Almost entirely because of the SUV, the overall fuel economy of North America's vehicles, after two decades of improvement, had started to deteriorate, falling to a level not seen in more than twenty years. And this was happening despite mounting evidence that climate change was real and urgent, and that greenhouse gas emissions coming from SUVs were one of the fastest-growing parts of the problem.

One of the reasons it has been difficult to deal with the SUV situation is that the North American auto industry, and its workforce, has become so dependent on the SUV for its profitability. It's hard to imagine how the kinds of improvements the Union of Concerned Scientists has advocated for SUVs—making them lighter and more fuel-efficient while maintaining their basic size and style—would destroy their appeal, or even undermine Detroit's dominance in their market. But without legislated action by government, there's simply no incentive for automakers to bother adopting these sorts of changes. In fact, it would probably be to their disadvantage to do so if others weren't required to do the same. Given the current trend towards heavier vehicles, a lighter SUV would probably be seen as a vehicle for wimps, who presumably would feel practically naked with, say, only 5,000 pounds of steel wrapped around them when other SUV drivers were protected by 6,000 or more. Only through tougher fuel economy requirements can the foolishness of ever bigger vehicles be stopped.

Sadly, instead of taking action to deal with the out-of-control SUV situation, the Bush administration and Congress have taken steps that actually exacerbate it. In the spring of 2003, an especially generous new tax measure favouring SUVs was tucked into the administration's $350-billion tax-cut plan. The measure was actually a reworking of the 1984 tax measure that allowed the self-employed to deduct the cost of purchasing an SUV. But while the earlier measure had capped the total cost that could be deducted at $38,500, the maximum was now bumped all the way up to $100,000, ensuring that America's other taxpayers effectively subsidize those purchasing SUVs—provided that the SUVs weigh in at 6,000 pounds or more. "It's a loophole," a car dealer's ad unabashedly pointed out in *The Dallas Morning News*, "and this weekend we can show you how to make that loophole big enough to drive a fleet of trucks and sport utility vehicles through it!" Henceforth, then, lawyers, doctors, dentists, accountants, real estate agents and other self-employed business people were to be permitted to deduct the full cost of a hyper-luxurious SUV on the grounds that it was a work-related vehicle, like a pickup truck, and therefore useful for toting around—who knows?—heavy equipment, dental tools, golf clubs or perhaps just bags of money.

Furthermore, in July 2003 the Republican-dominated U.S. Senate rebuffed fresh attempts to tighten CAFE standards. Instead, they voted overwhelmingly in favour of a measure (the Bond-Levin amendment to the Senate Energy Bill) that will effectively block future attempts to raise fuel economy standards by making it necessary for regulators to take into consideration the impact tougher standards would have on "manufacturer competitiveness." So, it seems, the Detroit automakers' fears that they will be unable to compete in the global economy will be considered sufficient grounds to kill tougher regulations. And those fears, as we know, are almost limitless.

The sheltered workshop lives on.

CHAPTER 6

THE GREAT ANACONDA

People often snickered at Samuel Van Syckel in the booming oil town of Titusville, Pennsylvania, in the 1860s. At the local hotel, where he ate his meals with many of the other entrepreneurial adventurers who had flocked to the area after oil was discovered in 1859, Van Syckel got used to being mocked for having big ideas. Embarrassed by the taunts, he eventually took to coming and going by the hotel's back door and eating by himself.

Oil had become a much sought-after commodity in the early 1860s. Although the car was still a far-off concept, oil was making a splashy debut on world markets because it could be refined into kerosene and used in lamps, providing a cheaper and brighter light than the prevailing coal or whale oil, especially now that whales were becoming scarce. But if extracting oil from the ground and refining it into kerosene proved relatively easy, getting the product to market was another matter entirely.

The most immediate problem was moving the oil from the many little wells that dotted the valley around Titusville to the small creek that ran down to the Allegheny River, along which the oil could be transported by boat to the rail line at Pittsburgh. The distance from the little wells to the creek was not great (generally under ten miles), but the trip was difficult. The oil had to be put into barrels and hauled by horse-drawn wagon through fields and forests and over narrow, rough, makeshift roadways.

Long oil caravans with dozens of wagons snaked along these paths. Often the whole caravan was held up for hours after a horse or wagon wheel became stuck in a mudhole. The hauling was mostly done by local men and boys, equipped with farm wagons, who had spotted an opportunity to collect exorbitant rates, thereby becoming perhaps the first—although by no means the last—to make handsome profits by gaining a virtual stranglehold on the movement of oil.

The solution to this tyranny of the oil haulers (or teamsters, as they were called locally) was a pipeline. Various attempts were made, but given the problems of leaking, bursting and overcoming gravity, the pipeline remained, well, a pipe dream—until Samuel Van Syckel showed how it could be done. Using a few strategically placed pumps, he laid a narrow pipeline just beneath the ground from the oil well at Pithole all the way (four miles) to Miller's farm. And it worked. Astonishingly, eight barrels of oil could move through Van Syckel's pipe per hour. Suddenly the teamsters no longer had a hammerlock on the situation. Although they responded by trying to vandalize the pipe, they soon saw that their days commanding a lucrative cut from the oil trade were over. After that, none of the oilmen in Titusville laughed at Van Syckel, who was no longer eating by himself. He had just completed what turned out to be a revolution in the oil business.

Van Syckel continued to be an important figure in the early oil trade, alternatively making and losing money in the ups and downs of the business, and continuing to come up with innovative ideas. In 1876, he took out a patent for a process that would allow oil to be refined on a continuous basis, thereby increasing the efficiency of the simple refinery operations of the time. With some financial help from an investment partner, he had just begun the construction of a small refinery along these lines when he was approached at his new site by a representative of the Acme Oil Company, a large concern in the refining business. Right on the spot, the representative made him an offer: the company would pay Van Syckel a comfortable salary for life if he would abandon his refinery project and let Acme build it instead. This was in many ways a good offer, but Van Syckel loved the life of inventor and entrepreneur, and he had no interest in working for someone else. Without any hesitation, he gave the man from Acme a firm no and considered that the end of the matter.

But it wasn't. Not even slightly dissuaded, the Acme representative told Van Syckel that it would be pointless for him to build his refinery; if he did, he wouldn't be able to make any money. The man went on to explain bluntly that Van Syckel would be obliged to pay railway freight rates so high that his oil would be hopelessly uncompetitive—because Acme had special deals with the railways. Van Syckel had heard rumours of these sorts of secret deals between the railways and the big companies. Everybody in the oil business knew this kind of thing went on. But he was taken aback by how boldly and coldly the man laid the facts out before him. It was the beginning of Van Syckel's long and tortuous encounter with the Oil Trust, a powerful set of interests controlled by John D. Rockefeller, who would prove much tougher to outsmart than the teamsters of Titusville.

Two days later, Van Syckel was in a grand building in New York City, being greeted by one of the senior figures of the Oil Trust. Perhaps it was the more luxurious setting (Van Syckel was also staying in a fine New York hotel, at the company's expense), perhaps the cold realization that he could not make it on his own with the company against him, but Van Syckel listened attentively this time to the deal the gentleman was offering. It was not ungenerous. He would be given $10,000 right away and paid $125 a month (a salary that would allow for comfortable living) while the company was building his refinery and testing the continuous distillation process outlined in his patent. If it proved successful, he would be given a further $100,000—a large sum of money—for his refinery and the ownership of his patent. It wasn't exactly as he had wanted things to be, but it wasn't really a bad outcome. Sitting in the comfortable chair in the finely appointed office, anticipating a nice dinner back at the hotel, Van Syckel felt all right, if not exactly pleased, about what he was doing.

"Let us put what we have agreed upon in writing," he said.

The gentleman from the Oil Trust explained that the paperwork needed to be completed, and it would be ready for him to sign in Titusville the following Monday. So instead of a written agreement, the gentleman offered his hand and said Van Syckel could count on his word of honour.

The following Monday, an official at the company office in Titusville told him the paperwork wasn't quite ready. Perhaps that afternoon.

Delays and more delays. The only thing that wasn't delayed was the company's takeover, and demolition, of the small refinery that Van Syckel had begun constructing.

Van Syckel never got his money. And the company never built his refinery. In fact, it blocked his efforts to find others to help him build his refinery, discouraging prospective investors with the same kinds of ominous warnings it had used to intimidate Van Syckel. Eventually, he was able to use some of his ideas in other refineries, with successful results, but the Oil Trust bought out these refineries as well. After twelve years of frustration, humiliation and increasing financial difficulty, Van Syckel filed a lawsuit against the Oil Trust with the Supreme Court of New York.

The rest of the story may seem easy to guess: high-priced lawyers for the Oil Trust making mincemeat of Van Syckel's case, leaving him even more devastated and showing how the big money-eyed interests can easily swat aside a mere citizen who tries to assert his rights. But in reality, the case had a few surprising twists. To be sure, the Oil Trust had a legion of pricey lawyers. Interestingly, however, they didn't simply swat aside Van Syckel's allegations. Rather, on behalf of the company, they confessed to them. The story, as laid out above, was uncontested by the company, including the part about how company representatives had made specific verbal promises to Van Syckel, leading him to believe he would receive substantial financial compensation, which he did not receive. The lawyers agreed that all this was true. But they insisted that, without a written contract, the deal was not legally binding on the company and no damages should be paid. In a stunning line of argument that sounded almost like an endorsement of Van Syckel's suit, they argued that he was a tragic illustration of the difficulty routinely faced by talented inventors trying to develop their patents.

"Mr. Van Syckel," the Oil Trust's lawyers told the court, "is an instance of what it means to get out a patent, and deal in patents— in nine cases out of ten. He was an inventive man. He has got out a good many patents. No question they were meritorious patents. And what is the result? Poverty, a broken heart, an enfeebled intellect, and a struggle now for the means of subsistence by this lawsuit."

Many people, upon hearing this summary of the case by the company's own lawyers, might conclude that it simply highlights the unfairness of the company's behaviour. And that was what the jury concluded: it decided in Van Syckel's favour, accepting his evidence and concluding that the company had in fact entered into a legally binding contract that it did not live up to.

So there's a twist: justice did triumph. Except for one thing. The judge said that it would be "the wildest speculation and guesswork" for the jury to try to calculate damages. He then ordered the jurors to set damages at *six cents*, leaving Van Syckel poverty-stricken and broken-hearted, and the case looking very much like a classic example of how big moneyed interests do manage to easily swat aside a mere citizen trying to assert his rights.

———

More specifically, Van Syckel's case is a classic example of how the Oil Trust operated—employing a mix of enticement, threats, coercion, double-dealing, lying, cheating, bullying and ultimately using its massive financial resources to crush opponents.

The story of the rise of Standard Oil, the forerunner of Exxon Corporation, has been told many times. Certainly, the details were widely known in the late nineteenth century and early twentieth century, due to the many highly publicized court battles against the company. There were also a number of widely read popular accounts of Rockefeller's treachery, such as Henry Demarest Lloyd's scathing 1894 best-seller *Wealth against Commonwealth*, and Ida Tarbell's exhaustive and detailed 1906 blockbuster *The History of the Standard Oil Company*. It's probably not too strong to say that Rockefeller's Oil Trust was a major issue on the political landscape at the turn of the century, and was considered by many to exemplify everything that had gone terribly wrong with America, as giant corporate powerhouses replaced the citizen-run democracy mythologized by the founding fathers.

Public attitudes towards Rockefeller seem to have softened over the years, with his name now associated mostly with philanthropy. The best-known version of the Rockefeller saga is no longer the harsh accounts of Ida Tarbell or Henry Demarest Lloyd, but the more forgiving version in Daniel Yergin's Pulitzer Prize-winning *The Prize: The Epic Quest for Oil, Money and Power*. Yergin portrays Rockefeller in a light that is, ultimately, flattering—as a ruthless operator to be sure, but one whose hardball tactics were needed to whip the oil industry into shape. Writes Yergin,

> Rockefeller was in some ways the true embodiment of his age, Standard Oil was a merciless competitor that would "cut to kill" and he became the wealthiest of all. Yet, whereas many of the other robber barons amassed their wealth by speculation, stock

and financial manipulation, and outright fraud—cheating their stockholders—Rockefeller built his fortune by taking on a youthful, wild, unpredictable and unreliable industry, and relentlessly transforming it according to his own logic into a highly organized, far-flung business that *satisfied the basic hunger for light around the world* [italics added].

Wow, that's quite a mini-portrait. It starts off with Rockefeller being merciless and ends with him bringing light to a darkened world. Along the way, he is contrasted with robber barons who are said to have done really bad things, like committing fraud and financial manipulation, and cheating their own stockholders. The implication is that Rockefeller, by contrast, was an honourable fellow who was only guilty of playing a bit of hardball with a "youthful, wild, unpredictable and unreliable industry"—in other words, with an industry that badly needed a little discipline. So maybe Rockefeller wasn't such a bad guy after all? Perhaps a little tough and unfeeling, but in the end just doing what had to be done to bring light to the far recesses of the world.

Yergin never explains why cheating one's stockholders should be regarded as a dirtier deed than, say, cheating Van Syckel—and countless others like him. If there is one thing that stands out in the history of the rise of Standard Oil, it is how Rockefeller never hesitated to lie, cheat or toss aside any code of fair play or common decency—not to mention observance of the law—in his efforts to eliminate all competition. Nor does Yergin explore the question of whether the "youthful, wild, unpredictable and unreliable industry" couldn't have sorted things out just fine without one player squeezing out all competitors through the use of coercive monopoly power. What if, instead, the industry had sorted itself out through the method that we at least pay lip service to—competition—under which (theoretically, at least) a business succeeds by running a more efficient, innovative operation than its competitors. It's not at all clear that Rockefeller would have achieved the dominance he did if he'd competed honestly, without resorting to treachery. Furthermore, there's no reason to believe that a competitive industry with a number of players would have been any less effective in delivering light to the world. In fact, it's likely that the world would have had just as much illumination but at a more affordable price—a difference that would have significantly benefited a lot of people all over the world over the years.

There's no question that Rockefeller was a self-made man who embodied the legendary qualities of a genuine entrepreneur. Born in 1839, he grew up in rural New York, the son of a lumber trader who later moved the family to Ohio and reinvented himself as a "doctor" selling patent medicines. Ambitious to make something of himself, young Rockefeller set out on his own in his early teens, after only a few years of formal schooling. He worked first as a bookkeeper and accountant, living in a Cleveland boarding house. At the age of nineteen, already accustomed to making his way in the world, he teamed up with an acquaintance and set up a produce-shipping business, where his frugality, keen sense of numbers and determination to get the best of any bargain soon became evident. The business thrived, particularly after the Civil War broke out, and the army became an important customer, with an endless need for produce. By 1862—only three years after oil had been discovered in Titusville—Rockefeller and his partner were approached about investing in an oil refinery in Cleveland. They spotted this for the hot opportunity it was.

The investment proved timely and wise. Cleveland, with its excellent rail and water links, was well positioned to deliver refined oil to the rapidly expanding west of the country, and the city's refineries were quickly coming to rival those in the oil regions of Pennsylvania. Rockefeller immersed himself in the briskly growing business and was soon a partner in one of

Cleveland's largest refineries. By 1870 he had bought out his original partner, started up a second refinery, opened an oil-selling concern in New York and consolidated his operations into one firm called the Standard Oil Company.

All reports point to Rockefeller's exceptional business abilities—his grasp of the industry, his talent for eliminating waste and inefficiency, his relentless hard bargaining—but right from the early days the record shows that the heart of Rockefeller's strategy was the elimination of competitors. While his business skills enabled him to win a substantial chunk of the market, they weren't enough to attain the kind of dominance he sought. In fact, by 1870 the field was becoming increasingly crowded; Rockefeller and his partners weren't the only ones to spot the lucrative possibilities in selling a substance that provided something as basic as light. The result of all this competition was dramatically falling prices and profits, as more and more drillers and entrepreneurs flocked to the business and flooded the market with oil. In 1865, Rockefeller had received fifty-eight cents a gallon for his refined oil; five years later he was getting only twenty-six cents. Competition, which was bringing down the price of oil for the consumer, was doing serious damage to Rockefeller's profit margins. He set his mind to the task of eliminating competition.

He seems to have achieved his early dominance in the industry to a large extent by making secret deals with the railways. As early as the late 1860s, one of his major rivals discovered that Rockefeller's success had at least something to do with the fact that he was getting substantially lower railway rates by receiving rebates on the officially posted freight rates. When a competitor requested the same deal, he was inevitably turned down on the grounds that the volume of his shipments, while big, wasn't as big as Rockefeller's. This may sound reasonable except that, at the time, the railways were effectively the nation's highways, and there was a popular notion that the railroad companies had a responsibility to operate without discrimination. Whether or not they had a legal obligation to do so would become a subject of intense political and legal wrangling over the years.

By 1871, the attempt to use the railways to gain a stranglehold on the oil industry had become much more ambitious and far-reaching. Some of the largest refiners and shippers entered into a "combination" with the major railways, creating a phony entity called the South Improvement Company, through which they colluded to divide up markets amongst themselves, discriminate against competitors and set prices. Rockefeller was the dominant partner in the South Improvement Company.

Under the most infamous arrangement of the scheme, refiners in the combination would receive not only substantial rebates from the railways as before, but also *additional payments of funds collected from their competitors by the railways.* For instance, the official freight rate for moving a barrel of oil from Cleveland to New York was $2. Any member of the South Improvement Company would pay this rate when it moved oil and then receive a rebate of 50¢, reducing its cost to $1.50 a barrel. But a competitor would be obliged to pay the official $2 a barrel—and 50¢ of that would then be handed over to the South Improvement Company. Thus, the South Improvement Company would receive a rebate not only every time it shipped a barrel of oil but also every time its competitors shipped a barrel of oil. Needless to say, under these circumstances there wasn't much opportunity for a competitor to compete.

The anti-competitive goal of the project was clear. The contracts establishing the arrangement spelled out that each railroad had to co-operate as "far as it legally might to maintain the business of the South Improvement Company *against injury by competition,* and to lower or raise the gross rates of transportation for such times and to such extent as might be necessary *to overcome the competition* [italics added]." To this end, the railroads were obliged to provide full details to the South Improvement Company of all shipments of oil by its competitors, allowing the company to keep tabs on exactly what the competition was up to. (The railroads benefited from this arrangement because they were assured the business of the major oil shippers and refiners in on the scheme; this gave them an advantage over competing railroad companies.)

Rockefeller approached the task of eliminating competition in a systematic fashion. Once the South Improvement scheme was fully in place, in January 1872, he paid a personal visit to each of the twenty-six refineries in Cleveland not affiliated with the South Improvement Company and explained to them how he intended the new system to work. The old system, under which Standard Oil had simply been receiving rebates, had been demoralizing enough for them over the previous few years. Now this.

Rockefeller minced no words in these personal visits to his fellow Cleveland refiners, making clear he intended to destroy their ability to operate as independent businesses. "You see, this scheme is bound to work. It means an absolute control by us of the oil business. There is no chance for anyone outside," he calmly explained to his competitors.

Having established the hopelessness of competing, Rockefeller then went on to offer his victims something. "We are going to give everybody a chance to come in. You are to turn over your refinery to my appraisers, and I will give you Standard Oil Company stock or cash, as you prefer, for the value we put upon it. I advise you to take the stock. It will be good for you."

Many of the Cleveland refiners were successful businessmen with resources and connections, not just struggling inventors like Van Syckel. One was Robert Hanna, the uncle of the future financial tycoon and political power-broker Mark Hanna. Robert Hanna and his partners went to see Peter H. Watson, the president of the South Improvement Company and a senior officer with the Lakeshore Railroad. After visiting Watson, they realized that Rockefeller's scheme really did have the backing of the railways, making resistance futile. Reluctantly, they handed over their refinery, for which they had paid $75,000, and received $45,000 from Standard Oil.

Within three months, twenty-one of the once-flourishing twenty-six Cleveland refineries had disappeared, absorbed in one way or another into what was becoming the Standard Oil empire.

———

Back in Titusville, Pa., reports and rumours had been trickling in about the mysterious goings-on in Cleveland. The people of Titusville and the other oil towns of northwestern Pennsylvania had long considered themselves locked in a battle of sorts with Cleveland, which increasingly threatened to replace the Pennsylvania oil region as the country's major oil centre. So news that many of the big Cleveland refineries were suddenly going under wasn't necessarily unwelcome—except that the Titusville crowd understood the dynamics of the industry well enough to realize that some sort of perfidy was probably involved. Those suspicions were only heightened when reports came in of a new entity called the South Improvement Company, which was said to be connected in some way to the Standard Oil Company. There were also rumours that a hike in railway freight rates was coming.

All this sounded ominous but was not particularly surprising—or unusually discouraging. The thousands who had flocked to the Pennsylvania oil region were a hardy lot who were used to the ups and downs of this rather exciting, exasperating new business; drilling for oil could produce a fabulous gusher or, more likely, just another dry hole. Certainly, there was much that seemed beyond one's control—and not just the luck of whether one's bit of leased land happened to lie over a rich bed of oil. More infuriating were the railways, in whose hands one was not only powerless but also, it seemed, the victim of capricious behaviour.

It was not uncommon to hear people in the oil region speak defiantly about building their own railroad.

The sense of victimhood had helped create a camaraderie and defiant spirit that permeated the oil communities of the area. Out of a virtual wilderness, and against considerable odds, people who had gone there penniless had managed to build thriving little towns, about which they felt considerable pride. Titusville, just a hamlet in the path of lumber traders in 1859, had been transformed into a city of ten thousand people, with stores, churches, schools (where students were assumed to be on their way to college), two newspapers and even, implausibly, an opera house where notable opera singers of the day performed.

There were dozens of thriving oil towns like this throughout the U.S. northeast and even in Canada. In southwestern Ontario, oil was discovered in Oil Springs in 1857 (two years before the Titusville find) and in Petrolia in 1861. Petrolia quickly turned into a boom town to rival anything in the Pennsylvania region. It had the highest per capita income in Canada, boasted impressive public buildings and contained some lovely Victorian homes. And while the people of Titusville often vowed to build their own railroad, the people of Petrolia actually did so—a venture that ended up paying for itself in the first year of operation.

The oil boom brought with it more than opera houses and stately homes. In the Pennsylvania region it also attracted some brawling, low-life elements drawn to the frontier atmosphere of the early oil towns. There were whisky peddlers and women of loose reputation, and it was not uncommon to see flatbed boats trailing little barges along the Allegheny River, delivering ample opportunity for wayward behaviour before moving on to the next town in the morning. The town of Pithole, Pa., was said in the early days to have had some fifty "free and easy" establishments offering "as big a den of vice as the world has ever seen." But this sort of thing was increasingly frowned upon. Those who felt the need for saloons or dance halls were soon obliged to seek them out exclusively in Petroleum Center, an otherwise respectable town where a tolerant citizenry had come to terms with the need for one street devoted to such activities.

In some ways, the oil region of Pennsylvania could be said to represent some of the best characteristics of nineteenth-century America. The industrious little towns were imbued with vaunted American values like thriftiness, decency and self-reliance, and they also had a strong sense of community. There was a lively sort of democracy in the region as well. The two newspapers in Titusville were among the feistiest in the country, openly slamming any vested interests (a favourite being the railways) that threatened to interfere with the viability of the local oil business. The *Oil City Derrick* was a particularly sharp and sophisticated contrarian voice, even spewing out Latin phrases at times to make its point: "*Sic semper tyrannis, sic transit gloria* South Improvement Company," the *Derrick* sneered. There was also in the new communities a deep-rooted sense of fairness and political justice that went beyond simply defending local oil turf. At the opera house, enthusiastic crowds turned out to hear such well-known human rights advocates as Bishop Matthew Simpson, who had given the eulogy at Abraham Lincoln's funeral, and Wendell Phillips, a leading anti-slavery activist and early defender of women's rights and universal suffrage.

Thus, on February 26, 1872, when the local newspapers reported that the long-threatened railway freight hike for oil had arrived, the spirited communities of northwestern Pennsylvania were prepared for a fight. The hike was massive, more than doubling the existing rate, thereby threatening to wipe out entirely the profitability of the local industry. What inflamed sentiments even more was the fact, reported clearly in the papers, that anyone affiliated with the South Improvement Company was to be exempted from the hike. So the worst fears had come true: the mysterious new entity was deliberately discriminating against all its competitors, and was clearly intent on driving them right out of business, just as it was doing in Cleveland.

The next day, some three thousand people jammed the Titusville opera house, with the crowd overflowing onto the street, to protest the treachery. Carrying banners with slogans denouncing "the conspirators," they shouted and ranted and vowed to take action to defend themselves. Three days later, another massive, angry crowd converged at Oil City, twenty miles downcreek. In the weeks that followed, the oil business was virtually shut down as the local population rallied from town to town denouncing "the Great Anaconda" and calling for the "enemies of freedom of trade" to be "shunned by all honest men."

A strategy for fighting back quickly emerged. The producers formed themselves into the Petroleum Producers' Union, under the leadership of Captain William Hasson, a respected figure with deep roots in the area. The producers announced they were refusing to sell oil to anyone known to be connected with the South Improvement Company, or to move oil on any railway that was in on the conspiracy. (The big independent refineries in New York and Pittsburgh were still eager to take their oil.) The producers launched an investigation into exactly who was behind the so-called South Improvement Company, and names of known conspirators were printed on a daily blacklist at the top of the editorial page of the *Oil City Derrick*. The investigation also quickly uncovered the same outrageous railway rebate scam that Rockefeller had used on the Cleveland refiners.

The producers soon found there was considerable support for their cause outside the immediate region. They dispatched a committee to the state legislature demanding that the charter of the South Improvement Company be revoked—which it quickly was. Another committee was dispatched to Washington, where it met with great success. Congress agreed to an investigation, and

President Ulysses S. Grant himself met with the committee members for an hour. In that session, Grant told the delighted producers that the national government might have to take action to protect the public against monopolies.

So the resistance was proving surprisingly effective. By late March, only a month after the news of the freight rate hike had hit the Pennsylvania region, the railroads announced they were withdrawing from the scheme. And in Washington, as officials of the South Improvement Company were grilled by congressmen in public hearings, details of the nefarious scam filled the pages of newspapers across the country, prompting plenty of editorial outrage. With this sort of mounting pressure, the congressional committee quickly delivered its judgment: the South Improvement Company—which now no longer even had a legal charter—was "the most gigantic and daring conspiracy" ever seen in a free country. The Great Anaconda was on its knees. The South Improvement Company had been wrestled to the ground. Rockefeller had been held up to national ridicule and hostility.

But, for all the public action taken against him and all the public venom spewed out about him, the truth was that Rockefeller emerged from the "Oil Wars of 1872" richer and more powerful than he'd been before. He now owned almost all the refineries in Cleveland; nothing had been done to reverse that. The very public dismantling of the South Improvement Company didn't even seem to faze him. He calmly, quietly went about operating as he had before. Very soon he had worked out new rebate deals with the railways.

A year later, in May 1873, Rockefeller himself appeared on the streets of Titusville, going almost door to door with a few associates, openly promoting a scheme for a new sort of combination, and arguing that the end of competition would be in everyone's best interest. The embittered people of Titusville and neighbouring towns answered with a firm no. But three months later Rockefeller reappeared, this time with the news that about four-fifths of the national refining business had formed something called the National Refiners' Association. Rockefeller was its president.

The Petroleum Producers' Union swung back into action, more determined than ever to slay the reincarnated Anaconda. Rallying once again under the leadership of the resolute Captain Hasson, they created an agency that would purchase oil from them as individual producers so that they wouldn't be at the mercy of Rockefeller's Refiners' Association. And they immediately declared a six-month ban on drilling, in order to starve Rockefeller of oil.

They held together and by early November seemed to have Rockefeller on the ropes again. On November 8, he was ready to recognize their new agency, and came to them with an offer. His association was prepared, he said, to sign a contract immediately to buy oil at the very generous price of $4.75 a barrel. He also pledged that, as long as this deal was in place, he would take no rebates from the railways.

To Hasson, the answer was obvious: No! Rockefeller and his association couldn't be trusted. But others weren't so adamant. The strain of the drilling ban was wearing down many of the producers; they wondered how much longer they could hold out. There was already evidence that some producers had been selling oil to Rockefeller's association on the side. And there were reports of wildcat drilling operations that had yielded substantial new oil supplies, weakening the effectiveness of the boycott. In fact, there was a lot more oil around than the producers cared to admit. In light of that fact, Rockefeller's offer to pay $4.75 per barrel—well above the going rate—was particularly tempting.

At a fateful meeting in Oil City on December 12, the producers were sharply split over what to do, and the fighting among the factions was so intense and rancorous that the *Derrick* couldn't bear to report it. In the end, the forces wanting to deal with Rockefeller prevailed.

It was just a month later, after the producers had shipped about fifty thousand barrels of oil under the new contract, that Rockefeller's association abruptly announced that the contract was null and void. No more oil would be bought at the elevated $4.75-a-barrel price.

It was later learned that Rockefeller was still receiving rebates from the railways. In truth, he had never stopped receiving them.

———

The media quickly dubbed the international protest movement that emerged in the late 1990s the "anti-globalization" movement, but this made little sense. There is nothing remotely anti-global about this worldwide movement, with its protests in far-flung corners of the globe and its annual gathering of thousands in Pôrto Alegre, Brazil. Probably more than any phenomenon of recent years, this movement has been facilitated by that ultimate global connecting system, the Internet, through which activists all over the world can communicate with each other effortlessly and cheaply. The notion that the millions of people around the globe who loosely make up this movement are against the idea of connecting the world more closely is, quite simply, absurd.

What defines this movement is not an opposition to global connectedness but an opposition to *corporate control*. The fact that corporate control is global in reach perhaps accounts for part of the confusion. The other part may well be deliberate obfuscation: those who feel threatened by the protest movement and what it stands for like to characterize it as anti-global and thereby portray its adherents as anti-modern, as the new Luddites. (As we know, the real new Luddites are more likely to be found in the boardrooms of oil or automobile companies.) It's not surprising that a worldwide popular movement to resist corporate control should arise at the end of the 1990s, after almost two decades during which the dominant ideology has been the abandonment of all attempts to regulate or restrict corporate power. It's interesting to note that this powerful new anti-corporate-control movement has roots that can easily be traced back to the late nineteenth century, when the modern mega-corporation first emerged in all its dominating, swaggering, world-controlling form.

If there is one example that nicely captures the phenomenon of the rise of the new corporate behemoth, it is Rockefeller's Standard Oil Company. The story told above is just a snippet of the four-decade-long tale of the consolidation of the Rockefeller empire, as it succeeded in achieving an astonishingly tight grip on the North American oil industry by the first decade of the twentieth century. (In Canada, a group of entrepreneurs hoping to keep Canada's oil fields beyond Rockefeller's grasp had created the Imperial Oil Company in 1880. But, unable to withstand his familiar tricks for snuffing out competition, Imperial became a branch of the Rockefeller empire in the 1890s.)

What is striking is how differently the public reacted to surging corporate power and its abuses back then. Aside from those in the protest movement, our age seems to have largely resigned itself to being under the thumb of powerful corporations. Even corporate fraud scandals on the scale of Enron and Global Crossing, and reports that corporate manipulations caused a summer of blackouts in California, seem to do little to rouse the public to demand some sort of meaningful restraint on corporate power. Similarly, news that George W. Bush raised some $200 million from the corporate sector for his re-election bid, clearly suggesting that the corporate world enjoys huge influence on his presidency, was greeted largely with equanimity.

The public mood was very different back in the late nineteenth century and early twentieth century, when full-throttled corporate power was first asserting itself. The resistance to the rising corporate juggernaut was also full-throttled and widespread, taking the form of popular protest movements. One that was particularly influential was "Progressivism." Its broad support across the political spectrum nationwide compelled both Republican president Theodore Roosevelt and Democratic president Woodrow Wilson to at least appear to address some of its concerns. The reformist zeal of Progressivism was fuelled by the rise of "muckraking" journalists who documented the cozy relationship between government and the new corporate elite. Woodrow Wilson stated the problem bluntly in his successful 1912 bid for the presidency: "If there are men in this country big enough to own the government, they are going to own it."

The rise of corporate power marked a striking change in the nature of American society. Up until about 1870, America was largely a rural-based, small-town society; local elites were made up mostly of merchants, small manufacturers and professionals who enjoyed considerable power, influence and prestige. This changed significantly after the Civil War, with the post-war reconstruction boom and the rapid growth of big cities. The "corporation" emerged and quickly became the dominant form of enterprise. With no real constraints on corporate growth or regulation of corporate activities, a small number of corporations soon became dominant in their economic spheres, assuming what amounted to monopoly power over key markets like oil, steel, beef and sugar. This was capitalism at its rawest and crudest, its rapacious nature captured and frozen in time forever in the board game Monopoly.

Along with this transformation came the rise of a fantastically rich new elite, the likes of which had never been seen before in America, where great fortunes had been associated with the landed gentry left behind in the Old World. In the 1840s, there had been fewer than 20 millionaires in the United States; by the early 1890s, their ranks had swollen to more than 4,000. There were even 120 Americans with fortunes over $10 million. Where all this was leading was a question raised with some alarm in a widely discussed 1891 magazine article titled "The Coming Billionaire."

For many Americans, the rise of a homegrown aristocracy threatened to significantly alter American democracy, changing the nature of their society and jeopardizing the individualist tradition. To these sorts of charges Rockefeller replied, in effect: Get used to it. He insisted that the type of corporate concentration he had pioneered was an essential step in the evolution of business. "It has revolutionized the way of doing business all over the world . . . " Rockefeller said. "The day of combination is here to stay. Individualism has gone, never to return."

To Progressives, these were fighting words. What was at stake was the very survival of the American way of life, which revolved around concepts like individual effort, enterprise, thrift, hard work, perseverance—concepts that were seen as essential to the pioneer experience. Nothing was more central to the mythology about America than the notion that it was a land of opportunity, that character and hard work, not family background, detemined success.

So it wasn't just that monopolistic corporate power pushed prices higher (which it did). More importantly, this power threatened individual opportunity, striking a potentially lethal blow at a central principle of America. This became a major theme in the political discourse of the time. "There is a sense in which in our day the individual has been submerged," wrote Woodrow Wilson, who became one of the most prominent articulators of the Progressive critique of corporate concentration and power. "There was a time when corporations played a very minor part in our business affairs, but now they play the chief part, and most men are the servants of corporations . . . Anything that depresses, anything that makes the organization greater than the man, anything that blocks, discourages, dismays the humble man, is against the principles of progress."

In reality, Wilson didn't go after the big trusts with the zeal that some of his words suggest. Still, the main point here is that concerns about the dangers of corporate power were very much in the mainstream and were considered central to the health of the nation's democracy. Wilson characterized the struggle against the trusts as nothing less than "a second struggle for emancipation." And he described this corporate power as something that had to be resisted on many fronts. He lamented, for instance, the fact that Americans were increasingly being reduced to mere employees of big corporations. In a line that is striking to read today, Wilson insisted that, if future generations found themselves "in a country where they must be employees or nothing, then they will see an America such as the founders of this Republic would have wept to think of."

Certainly the notion of the individual and individual rights lay at the very centre of the American democratic ideal that the founding fathers espoused (even if they didn't think of extending it to women and black Americans). The rise of awesome corporate power in the late nineteenth century prompted Wilson and the Progressives of his day to worry that the mighty new force of the corporation was crushing the individual so completely as to make democracy meaningless and unrecognizable. Today, we don't hear much about that sort of thing, about the dangers of corporate power overwhelming the "humble man [or woman]." Individual rights are still invoked, but now the mantle of individualism has largely become the terrain of the right and the corporate world. From the founding fathers and the Progressives, the torch of individualism seems to have been grabbed by the likes of Exxon's Lee Raymond, who fiercely champions the rights of the individual to churn just as much greenhouse gas into the atmosphere as he or she desires.

———

Nobody was under any illusion that Frank B. Kellogg was about to tell a happy tale when he rose to his feet in a New York City courtroom in August 1907 and said, "I shall tell you a story of the Standard Oil Company."

Kellogg was one of the top lawyers in the country, and he had been hand-picked by President Theodore Roosevelt for this titanic legal battle, which pitted the U.S. government against Rockefeller's sprawling oil interests. For four decades the Rockefeller empire had been systematically consolidating its control over the oil industry, and it seemed unstoppable, despite dozens of legal challenges, congressional and state investigations, and constant denunciations in the press and political campaign speeches. Now the full force of the federal government, at the behest of the president, was to come down against Rockefeller, in the form of an exhaustive antitrust prosecution that would probe every secret deal that had allowed Standard Oil to gain and maintain control over the nation's oil industry.

The case was massive, eventually filling 23 court volumes, with 1,371 exhibits and 14,495 pages of testimony from 444 witnesses, including the company's top officials and even Rockefeller himself. Now sixty-nine years old, somewhat gaunt in the face and wearing a white wig, Rockefeller faced Frank Kellogg's barrage of questions with studied indifference and evasiveness. With public attention riveted on the courtroom drama, Roosevelt sought to reinforce his presidential image as the crusading trust-buster when he publicly declared: "Every measure for honesty in business that has been passed in the last six years has been opposed by these men."

After two years of proceedings, the U.S. Circuit Court delivered a guilty verdict—only to have it promptly appealed by the Rockefeller lawyers. It wasn't until another two years had passed, in May 1911, that the chief justice of the Supreme Court, Edward White, delivered the Court's unanimous verdict: guilty. "No disinterested mind," the chief justice told the packed, silent courtroom, with the crowd hanging on every word in the 20,000-word judgment, "can survey the period in question without being irresistibly drawn to the conclusion that [Rockefeller's] very genius for commercial development and organization . . . soon begat an intent and purpose to exclude others . . . from their right to trade and thus accomplish the mastery which was the end in view." The Court gave Standard Oil six months to divest itself of all its subsidiaries.

It was a stunning victory. With the White House, the Supreme Court and public opinion all lined up against Standard Oil, Rockefeller seemed finally to have met his match.

The massive empire was quickly dismembered, broken up into a number of smaller—but still huge—corporations. The original holding company, Standard Oil of New Jersey, which represented almost half the net value of the conglomerate, was kept intact, and went on to retain its position as the dominant player in the oil business, eventually taking on the name Exxon. The other divvied-up parts of the conglomerate are also household names today: Standard Oil of New York later became Mobil, Standard Oil of California (or SoCal) became Chevron, Standard Oil of Indiana became Amoco, Standard Oil of Ohio became the American arm of British Petroleum, Continental Oil became Conoco, Atlantic became ARCO. "Public opinion and the American political system had forced competition back into the transportation, refining and marketing of oil," concludes Daniel Yergin, apparently satisfied that the dragon had been slain—even though, as he points out, the new companies operated more or less as they had before 1911, staying out of each other's turf and carrying on their old commercial relationships. The slain dragon still apparently had some life in it.

In fact, it turned out to have a great deal of life in it. While the impact of the Supreme Court ruling was undoubtedly significant, it would be wrong to see it as ushering in an age of meaningful competition in the oil industry and as curbing the economic and political clout of big oil interests. The companies that had formerly constituted Standard Oil, although technically broken up, continued to operate as a kind of inner community of companies dominating the oil industry, even without a centralized management. The ongoing role of the Rockefeller family itself can be glimpsed in a study prepared by the U.S. Securities and Exchange Commission in 1940, almost thirty years after the Supreme Court ruling. The study—the only investigation of its kind in corporate history—tracked the ownership of the largest corporations back to their "beneficial" owners (those who actually benefited from the stock, not just the "owners of record" listed in company documents). Although the stock of the big oil companies was widely held, the Securities and Exchange Commission concluded that the Rockefeller family enjoyed a degree of influence amounting to "working control" over Exxon (with 20.2 percent), Mobil (with 16.3 percent), Standard Oil of Indiana (with 11.3 percent) and Standard Oil of California (with 12.3 percent).

The oil industry remained a tightly knit club in which a very small number of players—particularly the group just mentioned—maintained a significant degree of control over the market. Key to this control was "an extensive network of intercorporate relationships achieved through joint ventures, intercorporate stockholdings and interlocking directorates," according to John M. Blair, who served for fourteen years as chief economist for the U.S. Senate's subcommittee on antitrust and monopoly. Blair noted that the problem of interlocking directorates—under which representatives of apparently competing companies sit together on another company's board—offered limitless opportunities for collusion and co-operation in restraint of trade. Blair's detailed chart of the interlocking directorships between the oil and banking industries in 1972 showed how often directors of the big oil companies came together on the corporate board of a financial institution to plan business strategies that ultimately affected all parties involved. These sorts of inter-relationships within the oil industry, as well as its interlocking connections to the banking industry, had grown substantially over time, Blair noted.

All this would be serious enough even if the impact had been confined to oil prices. But as we've seen, the oil industry was, and continues to be, very effective at influencing government on a number of important public policy issues. Thus, its immensely concentrated power has reverberated in the political as well as the economic realm. Many decades after the Supreme Court's 1911 ruling, prominent economist Harold Barnett described the U.S.-based oil industry, still dominated by the progeny of Rockefeller's Standard Oil, as "the greatest aggregation of effective and political industrial power which the world and nations have ever known." That assessment was made back in 1974, long before the political power of the oil industry reached its present, dizzying heights under the Bush administration.

Yet we seem to have all but lost interest in attempting to keep that power in check, or even keep track of it. This is certainly a switch from the days when monopolies were a burning political issue and corporate concentration was widely seen by the public and by government as a significant problem, even a threat to democracy. It was for this reason that the U.S. Congress, in 1914, supported passage of the Clayton Act, which was aimed at limiting the concentration of power through interlocking directorates. A House of Representatives report accompanying the bill noted: "The concentration of wealth, money and property in the United States under the control and in the hands of a few individuals or great corporations has grown to such an enormous extent that unless checked it will ultimately threaten the perpetuity of our institutions."

This concentration of ownership, and the power it delivers, hasn't gone away. On the contrary, it has intensified with the mergers and acquisitions of the past two decades. This is certainly true in the oil industry. The "Seven Sisters" that dominated the oil scene back in the 1970s are an even smaller family today. As economist Michael Tanzer notes: "Through incestuous marriages, the fabled 'Seven Sisters' have been consolidated into four mega-majors"— ExxonMobil, Royal Dutch/Shell, BP and ChevronTexaco. Yet even as the scope of the problem has grown, the concentration of corporate power has faded as an issue. Congress last dealt with the subject of interlocking directorates in 1914. The Securities and Exchange Commission hasn't probed to determine the real, "beneficial" ownership of the largest corporations since 1940. There hasn't been a thorough investigation of concentration in the petroleum industry since the 1950s, when the subject was exhaustively studied by the Federal Trade Commission and the Senate subcommittee on antitrust and monopoly. Even the more general subject of concentrated corporate power hasn't really been examined by Congress since its hearings on multinational corporations in 1974.

Those in Exxon's God pod today can relax in a way that Rockefeller, hounded by government, courts and the press, never would have thought possible. These days, the oil industry is spared attacks, or even gentle probes, from the government. The sweeping power wielded by the industry is effectively off limits as a political issue. No congressional committee or branch of government studies the concentration of power in the industry, or even mentions it. The major companies, once the subject of ongoing vigilance on the part of suspicious citizens and active governments, have become all but invisible, apparently benign.

Oil company owners? Why, they're just regular folks.

———

Rockefeller had managed, in only a few decades, to gain effective control over the North American oil market—and to amass untold riches for himself in the process. But that wasn't all he wanted. Increasingly, Rockefeller and the corporate empire he spawned began to cast covetous eyes at the rest of the world.

CHAPTER 7

HOW DID OUR OIL
GET UNDER THEIR SAND?

Bells started ringing almost immediately inside Ali Attiga's head.

It was 1967, and Attiga was a bright young Libyan economist. After being educated at the University of Wisconsin, he'd returned to his native country and by the late 1960s was heading up the research department at Libya's central bank. In that capacity he also had a seat on the Higher Petroleum Council, a panel that advised the Libyan government on matters relating to oil, something that was becoming increasingly important to the remote, underdeveloped North African desert kingdom.

Oil had been discovered there in the 1950s, and by the 1960s Libya was an oil exporter. There was every indication that a lot more oil lay under the desert sand, and the Libyan government was keen to see it developed. By 1967, some promising areas had been divided into lots, thirty-seven in all, and those lots were now being offered up for auction. All the major oil companies as well as dozens of independents had submitted bids, which had been passed on to the Higher Petroleum Council for review.

As Attiga and the other six members of the council looked over the bids, a striking fact was quickly evident: thirty-six of the lots had attracted no bids, while there were more than a hundred bids for one particular lot. "You didn't have to be very smart to see it," recalled Attiga in an interview in the summer of 2003. "There was one single lot that was sought after by everyone."

The potential of that one lot had become well known in oil circles through the grapevine. The lot had actually been explored some years earlier by Mobil, which had obtained a concession from the Libyan government. It wasn't long before Mobil's chief geologist had determined there was an enormous pool of oil there. But the geologist apparently hadn't seen fit to share this information with Mobil. So, assuming the site wasn't promising, Mobil had relinquished the concession. In the meantime, the geologist had left Mobil and joined a small but aggressive new independent company on the oil scene, California-based Occidental Petroleum. In the 1967 auction, Occidental was one of the bidders intent on grabbing that particular lot, and it knew better than anyone just how rich the field was. As Attiga and the others on the petroleum council looked over the bids, there was no missing Occidental's offer, which was written on sheepskin manuscripts and wrapped in red, black and green ribbons (the colours of Libya's flag). And it included a "signing bonus" of $30 million.

Attiga considered Occidental a particularly bad choice. The company was barely even in the oil business. It was owned by Armand Hammer, a flamboyant promoter who had made a fortune peddling treasures once owned by the czars of Russia. Hammer had recently decided to enter the oil business almost as a lark because, as he later explained, "an accountant friend pointed out that in my tax bracket it would not cost very much even if a venture lost money." Certainly, Hammer had no experience or understanding of the business, confessing, "I would not have known a barrel of oil if I fell into it."

Entrusting the development of such a key Libyan oil field to this character seemed like a bad idea to the members of the petroleum council. But ultimately the decision lay not with them but with the nation's doddering old monarch, King Idris. And the king was dazzled by the lavish Occidental bid; Hammer had even thrown in a solid gold chess set for the king as an additional signing bonus. The Libyan monarch was known for appreciating this sort of expensive trinket. As one top U.S. oil adviser later put it, "King Idris' regime was thought to be sound because it was corrupt." Being "sound" apparently meant being malleable, easily purchased. Western oilmen were used to getting their way with old King Idris. By the late 1960s, most of the Middle East's oil exporting countries were receiving ninety cents a barrel from the companies, but Libya was receiving only thirty cents a barrel, despite the fact that its oil was among the highest quality in the world and among the cheapest to bring to market, given its ideal location directly south of Italy on the Mediterranean Sea.

As Attiga contemplated the situation, he was struck by another thought. The intense interest shown by all the oil companies in that one lot made it clear that there was an abundant bed of oil there just waiting to be nudged out of the ground. Why did Libya need any of these foreigners to do that nudging, least of all a hustler who knew less about oil than dozy old King Idris himself? Why not do the job themselves, and ensure that the bulk of the profits were kept inside Libya, where they could do something to improve the desperately poor lives of the Libyan people? "I recommended that we should not give that lot to any company," Attiga recalls. "We should keep it out of the basket—and form a national oil company."

The king rejected the idea of a national oil company and awarded the much-sought-after lot to Occidental. Attiga and two other members of the petroleum council resigned in protest over this squandered opportunity.

The lot awarded to Occidental turned out to be one of the biggest oil deposits in the world, and incredibly easy to exploit. Occidental, for all its inexperience, had no trouble scooping the rich black bounty out from under the Libyan sand. The oil practically shot out of the ground when probed. "It was almost like a lake," says Attiga. "Normally oil is in porous sand. This was concentrated." Occidental quickly became the sixth-largest oil producing company in the world. "Occidental made its fortune in Libya," sighs Attiga, imagining what the desert kingdom could have done with all that money.

Armand Hammer went on to join John Paul Getty as one of the world's richest men. The dream of a Libyan national oil company would just have to wait.

––––––

It's perhaps hard for us in the West to grasp the sheer force of nationalistic feelings in the Middle East, and how much these feelings centre on oil. With virtually no other natural resources and a history of being under foreign domination, the people of the Middle East have come to see oil as their lifeline, as something key to any vision of an independent future. And these nationalistic feelings have inevitably brought them into conflict with powerful corporate forces in the West, which see the region's oil as, if not their birthright, then at least their legitimately claimed property. This Western attitude was captured succinctly in a clever slogan that appeared on placards at an anti-war demonstration in Washington in the spring of 2003: "How did our oil get under their sand?"

One of the reasons that we in the West seem to have trouble grasping the extent of the nationalistic feelings about oil in the Middle East is that we seem to have little of these sorts of urges about our own oil. At least in Canada, the idea of adopting a nationalistic approach to our oil reserves has been actively discouraged, virtually banned from national discourse in recent years. It's a bit of a digression here, but let me just quickly note that Canada's brief foray into developing a more nationalistic approach to our energy resources was quickly halted in the 1970s, mostly because it ran into the morass of federal-provincial wrangling. Since virtually all of Canada's oil is located in Alberta, the attempt to impose a national energy policy in the interest of all Canadians was seen in Alberta as a power grab by the rest of the country to claim Albertan resources. This was, perhaps, not surprising.

What is surprising, however, is how little sentiment there is in Alberta for *Alberta* to assert more control over and reap bigger rewards from its own oil. The province has been unable or unwilling in recent years to drive a really good deal for itself with the international oil companies—the kind of deal that other similarly positioned oil-producing regions have managed to win. Edmonton economist and consultant Mark Anielski notes that Alberta collects a considerably smaller royalty on its oil and gas than does Norway or Alaska. For that matter, the government of Ralph Klein actually collects a much smaller royalty than did the regime of former Alberta Conservative premier Peter Lougheed, who forced the industry to pay significant royalties. Those royalties were invested in building the province's infrastructure and in the Alberta Heritage Trust Fund, which by 1986 had grown to $12 billion (Cdn.)—roughly where it stands today.

Anielski estimates that if the Klein government had insisted on collecting the same share of oil and gas profits that Norway does, the province would have collected at least $50 billion more (Cdn.) over the last decade than it actually did. That could have meant a lot more for the Alberta heritage fund. (Norway's Petrofund, for example, has amassed more than $100 billion [Cdn.]) Instead, it seems, money that could have stayed in Alberta has ended up in corporations with head offices in Texas.

This sort of massive diversion of resources out of the country, while strangely uncontroversial in Canada, has long been highly contentious in the Middle East. By the early 1950s, the international oil companies had managed to effectively gain control over

Middle Eastern oil, and a desire to take that control back became a potent force throughout the region. It was a drama that was played out most vividly—and tragically—in Iran.

The story centres around the charismatic Mohammed Mossadegh, who, along with Gamal Abdel Nasser of Egypt, became a towering nationalist figure in the Middle East. With their inspiring visions and powerful oratorical skills, both men attracted huge, devoted followings. However, unlike Nasser, who was a heavy-handed and repressive leader, Mossadegh was a democrat with a deep commitment to the rule of law and a profound admiration for British parliamentary institutions. In fact, Mossadegh was at the head of a movement that pioneered democracy—with tenacity and great success—in Iran.

Effectively reduced to the status of a British protectorate after the First World War, Iran had long been mired in a power struggle between the British-backed monarch, or shah, and an elected body known as the Majlis. Mossadegh, born into a privileged, politically well-connected family and educated in Paris, had returned to Iran and become involved with politics at an early age, always championing the Majlis as the democratic voice of the people and challenging the autocratic, foreign-backed power of the shah. Mossadegh was repeatedly elected to the Majlis, where he was regarded as a figure of extraordinary eloquence and conviction.

One of the most volatile issues in Iranian politics from early on in the century was the enormously favourable deal that the shah had granted to a British oil concern, the Anglo-Iranian Oil Company. Actually, the name Anglo-Iranian (or Anglo-Persian as it was first called) was a misnomer; it should have been called the Anglo-Anglo Oil Company, as there was nothing remotely Iranian about it. It was entirely British owned (slightly more than half by the British government), and in 1901 it had been given a sixty-year concession for all of Iran's oil. As a result, massive revenues from

Iranian oil flowed into the British treasury. Without these revenues, British foreign secretary Ernest Bevin once noted, there would be "no hope of our being able to achieve the standard of living at which we are aiming in Great Britain."

That was nice for Britain, but it didn't leave much for Iran, which received only a small share of the revenue and had no voice whatsoever over the company's management, nor even the right to audit the company's books. Indeed, the British government and the company, headed by the imperious Sir William Fraser, apparently regarded Iran as little more than a colonial outpost to be exploited at their pleasure. The Iranian oil centre, Abadan, was almost a caricature of a colonial city. In the British section there were neatly trimmed lawns, gardens, swimming pools and a country club for the foreign elite. Nearby but out of view, the Iranian oil workers subsisted in abject squalor in a shantytown called Kaghazabad, where they lived in dwellings pieced together from rusty old oil drums hammered flat. There was no running water or electricity, or even paved roadways. The Iranians and the British lived utterly separate lives in Abadan. Buses and theatres—even if the Iranians had been able to afford them—were reserved exclusively for the British.

The mix of anti-British feeling and a desire for greater democracy came to a head in Iran in 1949. With members of the Majlis demanding that the concession of the Anglo-Iranian Oil Company be revoked, the shah, Mohammed Reza Pahlavi, son of the earlier shah, tampered with the national election in order to secure a more pliant legislative body. His blatant interference provoked huge demonstrations across the country, particularly in the capital, Tehran. Mossadegh called on those favouring fair elections to gather in front of his house, which they did—thousands of them. He led the noisy throng through the streets, up to the lawn of the royal palace, and refused to leave until the shah agreed to hold new, fair elections. After a vigil of three days and three nights, the shah agreed.

In the election that followed, Mossadegh and a group of his close associates were elected and became a powerful bloc pushing constantly for more democracy and more national control over the country's oil. With enormous popular support, expressed in frequent street rallies, they relentlessly advanced the oil issue. By the winter of 1951, a Majlis committee headed by Mossadegh recommended that the Anglo-Iranian Oil Company be nationalized. A vote was held in the legislative body on March 15. The shah and the British ambassador had strongly pressured the elected deputies to stay home. Despite the intense pressure, ninety-six deputies showed up—enough for a quorum—and every single one voted in favour of the nationalization. Five days later the matter was put to the Iranian Senate, where half the members had been appointed by the shah. Even here, the vote in favour of nationalization was unanimous. The following month, a motion passed in the Majlis making Mossadegh prime minister—a position he agreed to accept only if the legislative body agreed to nationalize Anglo-Iranian. Once again, it was a slam dunk: the motion passed unanimously.

The British were flabbergasted as the Iranians, under Mossadegh, proceeded with the nationalization. It was regarded by the British as an outright theft of British property and even an affront to Britain's honour. "Our authority throughout the Middle East has been violently shaken," said British foreign secretary Sir Anthony Eden. U.S. secretary of state Dean Acheson later summed up the British attitude—and apparently his own—when he described the British as being upset by "the insolent defiance of decency, legality and reason by a group of wild men in Iran who proposed to despoil Britain."

In his account of the Iranian nationalization, Daniel Yergin, once again showing his pro-industry sympathy, depicts the British departure from Abadan as a sad, almost poignant development. He describes how the oilmen and their families gathered in front of their country club and how the parson locked their little church "that housed the history of this island community—'the records of those who had been born, baptized, married or had died, in Abadan.'" Then, as the oil expatriates pulled out to sea on a British cruiser and the band played "Colonel Bogey's March," Yergin tells us that the "passengers broke into song, a great chorus under the hot sun" and "[w]ith this musical burst of defiance, Britain bade goodbye to its largest single overseas enterprise." In Yergin's telling, somehow the Iranians have become the heavies for taking control of their own oil while the British have become helpless victims, singing out in defiance after being driven from their homes. Let's not forget that these were the same British who wouldn't allow Iranians to travel on their buses or visit their theatres.

The international oil companies quickly swung into action, collectively coming to the defense of one of their own. An assault on Anglo-Iranian was seen as an assault on the international oil order, on the sanctity of the rights of oil companies to the region's oil reserves. ("First they came for Anglo-Iranian and I did nothing, then they came for Exxon . . .") The major international oil companies thus co-operated in imposing a worldwide boycott on nationalized Iranian oil. The British and U.S. governments backed the embargo, and Washington pressured independent American oil companies to respect the boycott and refuse to enter into any contracts for developing Iranian oil. In May 1951, the U.S. State department issued a press release reporting that U.S. oil companies had indicated "they would not, in the face of unilateral action by Iran against the British company, be willing to undertake operations in that country."

The boycott succeeded in cutting off Iranian oil from world markets, devastating the Iranian economy in the process. "The embargo . . . was very effective due to the co-operation of the eight major oil companies," observed oil analyst Zuhayr Mikdashi. "This embargo had, as intended, a punitive effect on

Persia's economy." Iran's oil exports dropped from $400 million in 1950 to less than $2 million in 1952.

Even so, the boycott failed to bring Mossadegh's government to its knees. In fact, Mossadegh's popularity with the Iranian people only seemed to grow. And throughout the Middle East he quickly became a popular symbol of defiance of British power. On a trip to Cairo in the fall of 1951, he signed a friendship treaty in which Iran and Egypt agreed to "together demolish British imperialism." (This united stand was particularly striking given the fact that Iranians are Persian while Egyptians are Arab; they speak different languages, although they share the Islamic religion.)

With the boycott failing to force Mossadegh to back down, the British contemplated further action. British defense minister Emmanual Shinwell warned of the danger of allowing the Iranians to nationalize their oil industry with impunity: "If Persia were allowed to get away with it, Egypt and other Middle Eastern countries would be encouraged to think they could try things on." Prime Minister Winston Churchill, a strong supporter of British imperial power, increasingly became convinced that Mossadegh had to be stopped, possibly even overthrown. "Holding the line against Third World nationalism was one of [Churchill's] lifelong crusades, and in the sunset of his career he was determined to make a last stand," notes Stephen Kinzer, a *New York Times* correspondent and author of *All the Shah's Men*.

Washington had originally been reluctant to support a coup in Iran, but that changed with the end of the Democratic administration of Harry Truman and the election in 1952 of Republican Dwight Eisenhower, who was inclined to see Iran as a potential battleground between the U.S. and the Soviets. Kinzer reports that on June 14, 1953, Eisenhower was briefed, in very broadbrush terms, about CIA plans to overthrow Mossadegh. "That was all Eisenhower needed, and he gave his blessing. Around the same time Churchill gave his own secret—and much more enthusiastic—approval." The CIA sent senior operative Kermit Roosevelt, grandson of President Theodore Roosevelt, to Tehran. Kermit Roosevelt spent a week meeting secretly with the shah to win his support for the coup. Kinzer describes their clandestine meetings and how the shah initially was reluctant to endorse the coup out of fear it might fail. In the face of continued pressure from Roosevelt, however, the shah came around to the U.S. position.

In the end, the coup succeeded. Iran's democratically elected government was brought down, and the enormously popular Mossadegh was arrested and tried by a military tribunal, imprisoned for three years and then placed under house arrest until his death in 1967. The shah, who had briefly fled in panic during the coup, was returned to his throne with greatly extended powers— and the full support of the U.S. government. Thus began an Iranian reign of terror that was probably equal to anything Saddam Hussein later got up to in Iraq. (As for Kermit Roosevelt—he was awarded the U.S. National Security Medal by Eisenhower at a secret ceremony. After leaving the CIA, he spent six years working for Gulf Oil.)

The U.S.-led coup against Mossadegh became a defining moment in the Middle East. To some extent it served as a dispiriting lesson to those who longed for greater national control over their countries' oil industries. Clearly, the British and Americans regarded the region's oil as something they had a proprietary interest in, and they had shown they were prepared to actually overthrow a government that challenged that interest. At the same time, by exposing the apparently imperialistic aims of the British and Americans, the coup in Iran became a powerful rallying point for anti-Western nationalism in the region in the years to come. That resentment found its clearest expression in Nasser, the fiery young Egyptian colonel who took power after a popular uprising—more than a million people demonstrated in the streets of Cairo in January 1952—against the corrupt ruling monarchy and the British occupation forces that propped it up.

Nasser's appeal lay in his willingness to defy Western powers and his fierce advocacy of Arab sovereignty and unity, a message that he transmitted across the Middle East in his "Voice of the Arabs" radio broadcasts. He became an almost legendary figure, revered by Arabs who had long chafed under docile Arab leaders controlled by European powers. Nasser's stature among the masses was only heightened when he claimed Egyptian control over the Suez Canal, the vital conduit for moving oil from the Persian Gulf to Europe. An armed intervention by Britain, France and Israel ultimately failed, leaving the canal under Egyptian control. Nasser's victory in the 1956 Suez Crisis was celebrated throughout the Middle East as a symbol of defiance—and a rare victory over the West.

Rare indeed. For the most part, the predominantly Arab and Muslim nations of the Middle East suffered humiliating defeats at the hands of Israel and the West. The overthrow of Mossadegh was certainly one of the more crushing defeats of an attempt to assert sovereignty in the region, and it ushered in more than twenty-five years of ruthless dictatorship under the shah, who remained closely associated with Washington. With all democratic avenues of protest shut down in Iran, opposition gravitated towards the mosques, where large gatherings were still permitted and where a hardened form of resistance began to emerge. It was this resistance, in the form of Islamic fundamentalism, that eventually brought down the shah in the Iranian revolution of 1979 and sparked an anti-Western fundamentalist movement across the Islamic world. It is a bitter irony that Washington now talks about its desire to bring democracy to the Middle East. Democracy had already triumphed—against great odds—in Iran in the early 1950s, when it was overthrown by the U.S. government, creating consequences that still reverberate today.

At the time, however, the overthrow of Mossadegh and the installation of the pro-Western regime of the shah were considered a victory in Washington and London, as disobedient Middle Eastern nationalists had apparently been taught a lesson and Iranian oil was once again restored to Western control. With the shah's secret police systematically eliminating internal opposition, it was safe for the Anglo-Iranian Oil Company to return, for Abadan to once again be home to an expatriate oil community. So Anglo-Iranian (later British Petroleum or BP) went back to Iran, this time as part of a consortium that included the major U.S. oil companies as well.

The international oil order had been restored.

———

One of the most striking things about the international oil order that dominated the global economy from the 1920s on was the way it efficiently, comprehensively, systematically operated as a cartel.

As the potential of Middle East oil was discovered in the early part of the twentieth century, a few big international companies moved in to take control of the oil in this undeveloped desert area, which for centuries had been controlled by the Ottoman Empire and, after World War I, had been carved up into separate countries operating mostly under British influence. By the mid-1920s, an entity called the Iraq Petroleum Company (IPC) had been established to develop oil in the newly created country of Iraq. The IPC was actually a partnership owned by the five major European and American players in the oil business: Shell, Anglo-Iranian (BP), Exxon, Mobil and the French state-owned oil company Compagnie Française Pétrole. (A minor interest was

held by a fifth player, Calouste Gulbenkian, an Armenian entrepreneur who had first spotted the area's potential in 1904.)
Together, these five companies already dominated the worldwide
market for oil, and their intention in entering into the partnership
was to avoid any competitive rivalry from developing between
them as they brought this new region into the existing scheme of
things.

To this end, the partners set about figuring out how to co-
operate in an orderly fashion that would allow each of them to
preserve its dominant position in the worldwide market. Using
the old boundaries of the Ottoman Empire, they selected a huge
swath of land—which today includes all of Iraq, Saudi Arabia and
most of the Gulf sheikdoms—and agreed to develop it jointly
through their partnership and, crucially, to forsake any separate
development plans that might put them in a competitive situation
with each other. The agreement, thereafter known as the Red
Line Agreement (apparently because at one of their meetings
Gulbenkian drew a thick red pencil line around the target area on
a map), just happened to involve what turned out to be the richest oil fields in the world.

The companies' intention to avoid competition within their
own ranks was clear. But there was still the danger that other,
outside players would try to muscle in on oil inside the Red Line.
So the partners in IPC moved quickly to buy up as many concessions as they could from the Iraqi government, and also to buy
control of any outside players who occasionally ventured into the
area in search of a concession. But they ran into a major snag
when a fairly significant player showed up and could not easily be
swallowed. The new player was Standard Oil of California
(SoCal), one of the pieces of the old Rockefeller empire. SoCal
was a relatively modest player compared with its big sister,
Exxon. SoCal had developed into a mostly regional company and
had spent a lot of money unsuccessfully exploring for oil around

the world. Its luck had changed, however, when it happened upon oil in the small Persian Gulf state of Bahrain in 1932.

Three of the big players in the IPC partnership—Shell, BP and Exxon—immediately spotted the enormous problem this threatened to open up: actual competition. An internal Exxon memo described the situation in its starkest terms: if not some-how incorporated into the existing system of global markets, SoCal "would be obliged to become competitive," and this would "adversely affect the price structure" in world oil markets.

Clearly, some kind of accommodation with SoCal was nec-essary. But before anything could be worked out, the problem got dramatically worse. In 1938 SoCal discovered oil in Saudi Arabia, an almost unimaginable prize for a relatively minor player. King Ibn Saud was pressing SoCal and its partner, Texaco, for quick development of Saudi reserves and SoCal and Texaco obliged. Keeping the genie in the oil bottle thereafter became a lot more difficult. By 1946, Saudi oil was making it onto world markets at 90 cents a barrel, giving it a competitive advantage over oil selling at the prevailing rate of $1.28 a barrel.

Exxon had a solution: it would become a partner in SoCal's Saudi Arabian venture so that Saudi oil could be marketed through Exxon's global outlets rather than being dumped on world markets at a lower price, thereby introducing reckless competition into an otherwise tightly controlled oil scene. There was a problem with this solution, however. Saudi Arabia fell within the boundaries of the Red Line, and the Red Line Agreement specified that the IPC partners could participate in ventures within the boundaries only if they did so *as a group*. But the Saudi king's dislike of all things British ruled out any involve-ment by BP or Shell. And the French and Gulbenkian were very resistant to any alteration of the Red Line Agreement. The con-troversy dragged on for six years, until eventually Exxon convinced the others that it was in everyone's best interest to avoid a price war. The Red Line Agreement was thus adjusted so that Exxon and its sidekick Mobil—without their IPC partners—could join the Saudi deal.

But SoCal (and Texaco) had to agree to let Exxon (and Mobil) in on the Saudi venture—and why on earth would they? They were sitting on the world's biggest oil prize, the scope of which was already well appreciated; they didn't need to knuckle under to Exxon, an oil giant that imperiously straddled the world market, anxious to snuff out competition wherever it surfaced. A faction on SoCal's board argued forcefully against letting Exxon in on the Saudi deal, insisting that SoCal and Texaco go it alone and become the supplier of the independent refiners and retailers operating in the U.S. market. "Our earnings in Arabia are tremendous and are going to be greater, deal or no deal [with Exxon]," noted SoCal director Robert C. Stoner. This view was keenly echoed by James MacPherson, head of SoCal's operations in Saudi Arabia, who urged the company to use its new-found "gold mine" to transform itself into a major independent force, not just a ball carrier for Exxon.

If SoCal were to let Exxon in, SoCal's share of the spoils would be significantly reduced and, more important, it would have to operate within the constraints of Exxon's world scheme. On the other hand, if SoCal and Texaco developed Saudi oil on their own, they would be free to become a major challenger, selling to the U.S. independent refiners and retailers, and eventually to become, as oil expert John M. Blair put it, "the largest supplier of oil in the world, able to beat any rival or combination of rivals in free and open competition for markets." There were obvious benefits for SoCal shareholders, but it's also interesting to speculate whether a powerful, independent SoCal, operating outside the cozy club dominating the oil scene, might actually have introduced some real competition, which could have benefited millions worldwide by lowering oil prices.

In the end, Exxon and Mobil were allowed into the Saudi venture. The world's biggest oil supply would be developed by an all-American team of companies: Exxon, Mobil, SoCal and Texaco, operating under an umbrella company called Aramco. Among other things, the deal effectively ensured Exxon's long-term domination of the international oil scene. Just why SoCal agreed to go along with the merger, given the persuasive arguments against it, is not clear. Yergin says it was because the merger was favoured by SoCal's president, Harry Collier, who, says Yergin, was known as "one of the 'Terrible Tycoons,'" in recognition of his wilfulness. Yergin also argues that letting Exxon in was the sensible option, since markets and financing would be assured. But was there any doubt that markets and financing would be available to a company with control over the biggest oil supply in the world?

Blair offers another possible explanation for SoCal's acquiescence to Exxon's overtures. He notes that the Rockefeller family interests still had enough stock in Exxon, Mobil and SoCal to retain effective working control in each of the three companies. So the Rockefeller family ultimately could decide what SoCal would do. And from the Rockefeller point of view, it was clear that nothing would be gained by having SoCal operating on its own, creating actual competition in the international oil market. Why would the Rockefeller family want to see companies they controlled go head to head in fierce competition? Much better to have them all co-operating nicely, sharing the spoils—as they had in the good old days of the South Improvement Company.

Maybe SoCal president Harry Collier, the so-called Terrible Tycoon, wasn't really a wilful tycoon at all, but just a guy following orders—from very high up.

DATE: September 1928
PLACE: Achnacarry Castle, west coast of Scotland
EVENT: International price-fixing and local grouse hunting
BYOS (Bring Your Own Servants)

If one were trying to imagine a scenario in which a bunch of corporate tycoons secretly collude to carve up worldwide markets
and fix prices amongst themselves, it would be hard to come up
with anything more fitting than the rendezvous at Achnacarry
Castle in the late summer of 1928. The heads of the three dominant international oil companies—Shell, BP and Exxon—met at
the remote early nineteenth-century baronial-style castle, nestled
among fifty thousand acres of superb hunting grounds in the
Scottish Highlands, to plot how to bring an end to a virulent bout
of price competition that had broken out in oil markets in the Far
East and threatened to spread beyond. If there was one thing
these oil barons shared, in addition to a taste for grouse hunting
and fine liquor, it was an aversion to competition. And the isolated
castle (offered for rent—without staff—through a local rental
agency) afforded them the perfect opportunity to work out a deal
amongst themselves for restraining competition—and to put it
down on paper without anyone in the outside world being the
wiser.

Despite great efforts to keep the meeting secret, however,
word of it leaked out. But that didn't stop the three men (Walter
Teagle of Exxon, John Cadman of BP and Sir Henri Deterding of
Shell) from hammering out an agreement in writing that set the
course for the international oil order for decades to come. The
written document, dated September 17, 1928, was an agreement
among the three dominant companies not to compete with each
other, but instead to set quotas in order to maintain their existing market shares, to co-operate in sharing facilities and to
prevent surplus production from disturbing prices. The broad

agreement of principles established amid the hunting and imbib-
ing at the castle was followed by three much more specific
agreements, drawn up over the next six years, which laid out
exactly how the companies would co-operate in running what
amounted to international, national and local cartels. In addition,
four more big players—Texaco, Gulf, Mobil and Atlantic—were
brought in, making the arrangements much more comprehensive
in their coverage of international markets.

The agreements are astonishing in their detail. Every aspect
of maintaining market share was set out with great specificity:
how quotas could be revised under different circumstances, and
what should happen when quotas were exceeded or when they
were underused, a practice known as "undertrading." (So, for
instance, it was specified that an undertrader's quota would be
reduced in the next accounting period by distributing to the
other participants "one-quarter of the amount by which his
undertrading exceeds his 5 percent margin.") There were even
rules to be followed in the area of corporate advertising. Types of
advertising that were to be "eliminated" or "reduced" included
road signs, billboards, newspaper ads, signs at dealers' garages,
and novelties such as cigarette lighters. More basically, it was
specified how prices were to be set (by mutual agreement) and
how outside competitors were to be dealt with (preferably by
acquiring them).

All this was overseen by a full-time secretariat with offices in
London and New York, with the mechanics of the day-to-day
implementation being left to what are unabashedly described in
the corporate documents as "local cartels." These bodies of local
oil company officials had their hands full studying all the rules
and ensuring local compliance with the anti-competition direc-
tives. The extent of their activities is captured by the fact
(uncovered by a Swedish parliamentary investigation) that in
1937 the oil companies' local Swedish cartel held a total of 55

meetings at which 897 subjects were discussed. (The following year, 656 subjects at 49 meetings, and the year after that, 776 subjects at 51 meetings.) Clearly, preventing free market forces from being unloosed in the oil industry was painstaking, detailed work.

The companies have long since acknowledged the Achnacarry agreement, but have insisted that it was terminated in the late 1930s. However, the U.S. Federal Trade Commission, which found massive evidence of the cartel arrangements in the companies' files during an investigation in the 1940s, found no evidence suggesting the arrangements had ever been terminated. John Blair points to evidence suggesting that the framework of corporate co-operation among the oil giants, reaching all the way to the top, still existed as late as the early 1970s. The case involved deliberations among the majors over whether they would continue to negotiate jointly with Libya and countries in the Middle East. A representative of Bunker Hunt International, a large independent operating in Libya, later testified at U.S. Senate hearings that at one point in the deliberations he was called to give his views of the matter to the majors' "London Policy Group." When it was decided that the question was too big to be resolved by the London group, he was asked to repeat his views in an international conference call put through to Mobil headquarters in New York, where he found himself addressing a "meeting of the chiefs," chaired by Exxon CEO Kenneth Jamieson.

———

Some analysts downplay how effectively the majors were able to operate as a cartel. While acknowledging that the Achnacarry agreement was an attempt to limit competition, Daniel Yergin, for instance, suggests that the attempt largely failed: "[T]he oil companies were no more successful in implementing their new agreement than they had been in keeping their meeting at

Achnacarry secret in the first place." It's true that there was some competition; there were some independents, particularly in the U.S. domestic market, where they were protected by domestic antitrust laws. There were also some independent producers, refiners and distributors elsewhere.

But what is striking is how, for many years, the majors effectively kept the independents marginal and unable to provide significant competition or pose any real challenge to the majors' dominance. The majors were for the most part able to unilaterally set prices and establish policies for the industry. "Ever since the Achnacarry agreement of 1928 to divide up world markets . . . the big companies maintained a relatively stable price for crude oil, primarily by controlling its worldwide production," observes economist Michael Tanzer. Iran, for instance, discovered just how closely the majors collaborated with each other when they organized and enforced a worldwide boycott of Iranian oil for two years after the Iranian nationalization. Clearly, the majors were operating as a unit and forcing the rest of the industry into line—and that boycott happened in the early 1950s, more than a decade after the majors claimed the Achnacarry agreement had been abandoned.

The cartel-like power of the majors was demonstrated by the way they were able to exercise control over the supply of oil on a worldwide basis, ensuring that only as much oil as they deemed desirable would make it onto world markets. Too much oil would bring the price down, which they obviously didn't want; on the other hand, too little oil would drive the price up, potentially creating a serious downturn in the world economy, which they also didn't want. Only by controlling the overall amount of oil available could the majors maintain their desired price. But controlling the worldwide supply of oil was no easy task. Oil was located in countries all over the world, and governments were obviously keen to have it developed as quickly as possible (and on as favourable terms as possible) to ensure maximum revenues. Yet, to an amazing extent, the majors were able to control supply, holding back the development of oil fields when it suited them to do so.

One of the most striking things about the story of oil is how much of it revolves around the majors *suppressing* the development of oil. This doesn't fit with the popular notion that there was a never-ending quest to find as much oil as possible and get it to market as quickly as possible. In fact, the majors were keen to avoid a glut that would force down the price. So they were often doing their best *not* to find oil—or at least to prevent governments from knowing how much they'd found, so as to avoid coming under pressure to produce it and sell it. Exxon vice-president Howard Page, upon learning that Exxon geologists had just discovered a ten-billion-barrel oil field in Oman, is reported to have replied: "Well, then, I'm absolutely sure that we don't want to go into it, and that settles it. I might put some money in if I was sure that we weren't going to get some oil, but not if we are going to get oil."

Iraq offers an interesting example of how the majors conspired to suppress oil—or "swallow" it, as oilmen sometimes put it. The richness of Iraq's reserves was understood early on, and the majors, as we've seen, were quick to gain control of them, but not to develop them. Operating through the Iraq Petroleum Company, the majors had succeeded in convincing the Iraqi government back in 1925 to grant IPC a concession giving it exclusive oil production rights for the entire country. But having gained this immense concession, IPC produced surprisingly little oil. By 1950, less than 1 per cent of the country's oil field areas had been developed, much to the frustration of the Iraqi government. Moreover, IPC had already discovered many more Iraqi oil wells, but the company went to great lengths to conceal information about these discoveries from the Iraqi government.

According to an intelligence report cited at U.S. Senate hearings in 1974, IPC had drilled and found wells in Iraq capable of producing fifty thousand barrels of oil a day, but the company "plugged these wells and did not classify them at all because the availability of such information would have made the companies' bargaining position with Iraq more troublesome."

IPC was up to the same sort of tricks in neighbouring Syria. The Syrian government began to wonder if IPC was actually interested in producing oil—a suspicion that appears to have been justified. In an internal company memo, IPC's general manager noted that the company was "drilling shallow holes on locations where there was no danger of striking oil." In another internal memo, he wrote: "We want to set up a convincing window dressing that we are actually working the concession."

If the majors weren't always keen to develop oil fields, they were always very keen to prevent others from developing them. So they reacted fiercely in 1961 when Iraq, angered at the slow pace of development, withdrew IPC's rights to all parts of the concession that the company had failed to develop—that is, 99.5 percent of it—and tried to attract new players to finally get some oil pumping out of the ground. IPC executives were furious, threatening legal action against Iraq and even enlisting the support of the U.S. government to discourage independents from taking Iraq up on its very interesting offer.

Washington eagerly took up the cause of the wounded majors, despite the fact that U.S. government experts advised that the Iraqis were within their rights. It wasn't as if Iraq had seized any IPC assets; it had simply withdrawn parts of the concession that IPC had failed, for more than thirty years, to develop. At worst this could be seen as a breach of contract (and a weak case at that), but it was not a matter for international law or the involvement of a foreign government. Andreas Lowenfeld, a legal adviser in the U.S. State Department, noted in a memo:

"IPC's legal remedies are few, and we have no firm legal basis for telling independent American companies—let alone foreign companies—to stay out of Iraq."

But Washington was not deterred. Averell Harriman, the influential undersecretary of state in the Kennedy administration, had other considerations in mind, explaining: "We could not wish governments, such as Iraq, to get the impression that American oil companies can be pushed around." To make sure Iraq got the message, the State Department actively intervened to discourage independents from showing any interest in Iraqi oil. A State Department memo from July 1964 reveals that the department had "interceded with all American companies which to its knowledge or belief have expressed interest in Iraq land concessions in order to deter them from making offers to GOI [the government of Iraq]." The pressure appears to have been successful. A State Department telegram signed by undersecretary George Ball that same month reports: "The firms to which department has spoken are Sinclair, Union Oil, Standard of Indiana, Continental, Marathon, Pauly, and Phillips. These companies have been responsive to Department's urgings and it is therefore incumbent on us to make same effort with any new American or American-affiliated company which appears to be entering Iraq picture."

Even without Washington's intervention, the independents were well aware that any willingness to take up Iraq's offer would land them in trouble with the majors. And given the extensive control the majors exercised over the industry and markets worldwide, this was not something that independents would risk idly. Even after a compromise agreement was reached between IPC and Iraq in 1965, independents were wary about making deals with Iraq for fear this might jeopardize their dealings elsewhere with the majors. So, for instance, when a small American independent was offered a drilling contract by Iraq in 1967, it felt the need to get what amounted to permission from the majors. According to a State Department memo, the assistant manager of Santa Fe Drilling reported that his company "carefully checked the attitude of Iraq Petroleum Company parents [BP, Shell, Exxon] before signing a contract . . . as it did not want to jeopardize other Santa Fe contracts with IPC parent majors."

In addition to being boycotted like this, Iraq was punished directly by the majors with further cuts to its already low level of production. The majors had long kept production levels down in countries they considered ill-behaved. Exxon's Howard Page admitted to a U.S. Senate commission that the majors punished uncooperative countries with production cuts while raising production elsewhere to reward good behaviour. Typically, Saudi Arabia and Iran under the shah were favoured with high production. Iraq, on the other hand, had been the perennial whipping boy of the international oil order, particularly after it withdrew most of IPC's concession in 1961. Notes oil analyst Pierre Terzian: "Iraq, which had dared to stand up to IPC . . . was treated even more harshly; production stagnated and exploration was brought to a complete standstill." Issam Al-Chalabi, the former oil minister of Iraq, says that this constant punishing of Iraq helps explain "why the production capabilities of Iraq are well below its potential."

If further evidence were needed that the oil majors operated as a cartel, it is provided by the manner in which, almost miraculously, they ensured that worldwide oil production grew at a very steady, consistent rate in the early post-war decades, thereby preventing either a glut or a shortage on global markets. This steady rate of growth worldwide might seem easily accomplished if production in each country grew at the same rate each year. But for a variety of political and logistical reasons, there was enormous variation from country to country. For instance, Iran's production disappeared totally during the embargo of the early

1950s, and then later was developed at a greatly accelerated pace to reward the shah's pro-Western behaviour, while production in Iraq, Kuwait and Venezuela proceeded much more slowly. Yet, as John Blair noted, "[S]omehow the major oil companies have been able to 'orchestrate' these and other aberrations into a smooth and uninterrupted upward trend in overall supply." Indeed, despite these widespread variations, the overall world output of oil just kept increasing year by year at an astonishingly consistent rate from 1950 to 1972: almost exactly 9.55 percent per year. If the rate of growth the majors were aiming for was 9.55 percent a year, it turns out that the *actual* rate of growth varied from this desired rate by less than 0.1 percent in any given year! If the majors weren't pulling strings to make this happen, it sure was an amazing coincidence.

———

The whole point of all this control was to enable the majors to set the price of oil and thereby secure enormous profits—something they succeeded in doing. "Historically, the largest profits in the oil industry have come from maintaining a monopolistic price for crude oil, one considerably higher than the average cost of production," notes Michael Tanzer. That was easily achieved in the Middle East, where production costs were extremely low. John Warder, the chairman of the majors' consortium in Iran, revealed just how cheaply Iranian oil could be produced when he gave a presentation to the Central Bank in Tehran in 1967. Ironically, Warder was complaining about what he considered Iran's high production costs—of 14 cents a barrel—which he attributed to a surplus of personnel, and which he insisted were making Iran an uncompetitive place to produce oil. By comparison, he noted, oil could be produced for a mere 8 or 9 cents a barrel in Saudi Arabia, and for only 6 cents a barrel in Kuwait. A barrel of oil was selling at the time for about $1.80.

It's important to note that the "competition" Warder referred to wasn't price competition, which would have benefited consumers. The competition was between nations to offer the cheapest production costs in order to attract the interest of the companies, which made the decisions about where production would take place. With the companies controlling the price and the nations competing to offer them the lowest costs, you could say that the companies were advantageously positioned.

It's not surprising that the profits enjoyed by oil companies operating in the Middle East were staggering—roughly triple or quadruple the rates of profit in other industries. The Iraq Petroleum Company enjoyed an average 56 percent of profit rate on net assets from 1952 to 1963. In Saudi Arabia, Aramco's profits averaged 57 percent. And in Iran, the consortium turned an average profit of 69 percent between 1955 and 1964. It seems that even those high Iranian labour costs Warder complained about didn't stop the company from achieving a staggering profit rate of almost 70 percent.

———

The cartel crafted by the major oil companies was in many ways an ingenious piece of work, far-reaching in conception, massive in scale, precise in detail. One can even admire what oil analyst Paul Frankel has described as the "workmanlike manner" in which it was put together. The fact that it prevailed, effectively allowing the major companies to control the world's access to the most vital resource for almost half a century, is a testament to the cleverness of its design. What nobody anticipated was that this structure, so carefully crafted by these enormously powerful companies, would be brought down rather quickly by an unsophisticated nomad with little experience of the world.

———

For all its corruption and backwardness, the regime of Libya's King Idris was the first to figure out how to avoid becoming yet another powerless country in the oil companies' harem, constantly vying for higher production levels but receiving them only at the whim of those running the companies. One way around this, the Libyans deduced early on, was to invite in lots of independents, which they did. Soon Libya had a large number of companies producing its oil, and the country found itself in a surprisingly powerful position—a position quite different from that of the other oil-producing nations. While the other nations had to compete against each other—essentially pleading with the majors to take oil from them, not from the next guy—the Libyans had an ace up their sleeve. They didn't need to rely on the majors. If the majors didn't want to produce their oil, the Libyans could get the independents to crank up *their* production. Since the independents weren't part of the companies' cartel, there was nothing to hold them back from producing as much oil as they could, as quickly as possible.

In fact, that was exactly what they did. The result was a flood of Libyan oil onto the market, much of it produced by the independents. This threatened to create a glut on world markets and force down the price. The independents were already undercutting the official "posted price" set by the majors, selling their oil for less. Inside the cartel, there was nothing short of panic.

The majors could, of course, prevent a glut by cutting back their own production throughout the Middle East, thereby reducing the total world supply. But such a move would enrage the Middle East countries, already frustrated by having to accept much lower production levels than they wanted. And there was always the risk that, if production levels went too low, countries like Saudi Arabia and Iran would break their concession deals with the majors and seek out independents themselves, or even consider nationalization.

Into this delicate situation strode Mu'ammer al Qaddafi, a young man from a nomadic desert family who had joined the Libyan army with the dream of doing what Nasser had done in Egypt. In September 1969, Qaddafi seized power in Libya, changing the dynamics of the oil world overnight. Whereas King Idris had been trying to negotiate a royalty increase from the companies of ten cents a barrel, Qaddafi promptly upped the demand to a forty-cent increase. The boisterous Qaddafi also signalled that he planned to create a new way of dealing with oil companies. He threatened to shut down the operations of all twenty-one companies operating in Libya if they didn't agree to the higher royalty.

Qaddafi's apparent willingness to temporarily give up oil revenues strengthened his playing hand. The increase he was demanding, although large by the standards of the time, was actually not unreasonable, particularly given the high quality of Libyan oil. This was acknowledged by James Akins, a petroleum expert in the U.S. State Department, who concluded that, on the basis of the industry's own formula, the forty-cent increase was, if anything, low. Akins sent reports to this effect to Exxon, Texaco, Mobil and BP. But the companies were used to calling the shots and promptly rejected Qaddafi's demand, making a counter-offer of a five-cent increase.

While Qaddafi no doubt seemed like a bush league player to them, in fact he understood the dynamics of the oil industry well, mostly because he was being advised by Abdullah Tariki, the sharp, well-informed former Saudi oil minister. With Tariki's assistance, Qaddafi developed the clever strategy of first applying pressure on the independents, since they had no other sources of oil to fall back on. Occidental, for instance, was a huge player in Libya, but only in Libya. If Libya cut back its production, Occidental would be unable to meet its contracts with its buyers. So Libya put the squeeze on Occidental. When Occidental refused to capitulate to Qaddafi's demands, Qaddafi abruptly ordered Occidental to slash its production by a massive 300,000 barrels a day. Two months later, when the company still hadn't capitulated, its production was cut by another 60,000 barrels.

Occidental president Armand Hammer was desperate to find an alternative source. His negotiations with the Libyans were going nowhere (it probably didn't help that he showed his distrust of the new regime by flying every night by private jet from Tripoli to his suite at the Ritz Hotel in Paris so he could make secure phone calls to company directors in Los Angeles). He also made a trip to New York to ask Exxon chairman Kenneth Jamieson to sell him—at cost—the 360,000 barrels of oil a day the company was now unable to pump in Libya. Without this backup supply, Hammer argued, Occidental would have no choice but to give in to Qaddafi's demand—and that would only increase the pressure on Exxon to do the same. Hammer had a point. But Jamieson was so unwilling to help an upstart independent—a company that had constantly sought to undercut Exxon's prices—that he refused Hammer's request.

Meanwhile, the Libyans refused to blink, despite the massive loss of oil revenues they were suffering. As Qaddafi noted: "People who have lived without oil for five thousand years can live without it again for a few years in order to attain their legitimate rights."

With its feet to the fire, Occidental agreed to Qaddafi's terms. Within a few months the rest of the companies operating in Libya followed suit, including Exxon. "Libya was the breaking point of the control the majors had on the supply of crude oil," says Ali Attiga, the Libyan economist. Libya's annual oil revenues shot up overnight from $330 million to $1.4 billion.

The rules of the oil world had just been rewritten. After decades of unquestioned dominance by the majors, it had come to this: a scraggly upstart of a nation had delivered a stunning blow to the world's most powerful companies, backed up by the world's most powerful governments. For the first time, the master had been effectively confronted and outsmarted; his aura of invincibility was shattered.

Inside the harem, things would never be the same again.

CHAPTER 8

THE HAREM TAKES ON THE SISTERS
THE RISE OF OPEC

In about a dozen oil-producing nations, Libya's boldness stirred feelings that had long festered but had found little public expression. With the Libyan breakthrough serving as a sign of what was possible, this assortment of largely undeveloped Third World countries would rise up in the early 1970s in an astonishingly successful way. Asserting themselves collectively as OPEC, they would confront the West with a forcefulness and effectiveness that was without precedent—and that briefly posed a challenge to the very underpinnings of the global capitalist system.

———

OPEC was the brainchild of two men, separated by thousands of miles of ocean, living in countries that, at the time, didn't even have diplomatic relations. While the gulf between Venezuela and Saudi Arabia in the 1950s was immense, the same sort of nationalist dreams came to inspire both Juan Pablo Pérez Alfonzo, a descendant of a venerable Spanish family in the Venezuelan capital of Caracas, and Abdullah Tariki, the son of a camel owner who organized caravans crossing the Saudi desert.

Oil—and resentment of American oil companies—had long been a central theme in the politics of Venezuela. As we saw in Chapter 4, the nation's oil wealth was controlled from early in the twentieth century by a small domestic elite with close ties to U.S.

oil companies. The transparency and extent of the corruption in this relationship provoked considerable opposition, which was severely dealt with by Venezuelan dictator Juan Vicente Gómez. Rebel leaders were sometimes hung up alive on butcher's hooks, and political prisoners, attached to balls and chains, were thrown into the ocean. (When Gómez died in 1935, some fourteen tons of bars, chains and iron balls were discovered in the notorious Puerto Cabello prison.)

Miraculously, Venezuelans persisted in organizing resistance. A campus protest movement at the central university in Caracas in the late 1920s later led to a fierce crackdown by the Gómez regime, with hundreds of students thrown in jail, including a fiery young law student called Rómulo Betancourt, who was to go on to play an important role in Venezuela's history. Inspired by the bravery of the movement's leaders, dozens more students demanded that they too be imprisoned with their comrades at the Puerto Cabello compound. Among those offering themselves up for voluntary imprisonment in this hellhole was Juan Pablo Pérez Alfonzo, a fellow law student who was to become a political collaborator and lifelong friend of Betancourt.

With their release from prison, Betancourt went into exile. The death of Gómez in 1935 brought a somewhat less brutal dictatorship to power in Venezuela, allowing Betancourt to return and begin underground political organizing. Along with Pérez Alfonzo and a few others, Betancourt established a pro-democracy political party that set about the risky task of organizing labour unions—for which Betancourt was briefly forced into exile again. By 1939, however, the political situation had eased enough for Betancourt and his party to take part in national elections, winning seats in the Chamber of Deputies—a parliament with limited power. Among those elected was Pérez Alfonzo, who, at Betancourt's urging, became the party's expert on oil, a role Pérez Alfonzo took on with great zeal.

Oil, and the politics that surrounded it, had only recently emerged as a subject that could be openly discussed in Venezuela. Although the government relied heavily on revenues from oil, it had done little to drive a hard deal with the companies. As a result, the companies paid very low royalty rates and there was no consistency in those rates; a small number of companies paid the top royalty rate (a mere 15 percent) while the vast majority paid far less. As public attention began to focus on the issue, the military government headed by General Isaías Medina Angarita introduced a bill to bring about a more orderly system of royalty rates. While it might seem that the government was getting tough with the oil companies, it wasn't. In fact, the bill had been drawn up by two American experts, one of them Herbert Hoover Jr., son of the former U.S. president.

Still, the bill offered some increase in revenue, so Betancourt's opposition party (called Acción Democrática, or AD) decided to abstain rather than vote against it. But Pérez Alfonzo used the vote as an opportunity to deliver a stirring speech against foreign oil interests. "The manner in which they have exploited the wealth belonging to the Venezuelan people . . . is a public and notorious fact . . . The Venezuelan nation, which is aware of, and deeply feels, the despoilment of what in equity and justice is its right, cannot be satisfied with its relations being simply adjusted to the oil concessionaires' convenience." Pérez Alfonzo insisted that the bill should have gone much farther, collecting compensation for the ridiculously low royalty rates of the past and imposing tougher terms on existing oil contracts to curb "the exorbitant profits" the companies were taking out of Venezuela. His bold words resonated with the country's middle class, and helped the AD build support for a broad-based political movement against the ruling oligarchy that had so enriched foreign companies.

In 1945, a Betancourt-led coup overthrew the Medina Angarita dictatorship and began an ambitious process of democratization.

Pérez Alfonzo was made responsible for oil affairs and soon took measures to force the oil companies to pay higher taxes—measures that increased national oil revenues fivefold within a couple of years. But Pérez Alfonzo's vision of the possibilities went beyond merely collecting more revenue. He laid out a far-reaching set of guidelines aimed at redefining the traditional colonial relationship between the oil companies and Venezuela. He called for the creation of a national oil company, which would be given priority for future exploration concessions and be empowered to sell Venezuelan oil to the world. He called for a freeze on concessions to foreign companies, construction of refineries in Venezuela, and drastic cuts in the domestic price of energy in order to encourage industrial and agricultural development. The new government also established an oil workers' trade union, which immediately pressed for and won a wage increase of more than 100 percent from the oil companies.

When the first general elections were held in December 1947, the AD swept to power with 80 percent of the vote. Now officially ensconced as oil minister in a democratically elected government, Pérez Alfonzo set about putting his ambitious agenda into place in earnest. In a move that sent shock waves through the international oil industry in 1948, he imposed a fifty-fifty profit-sharing arrangement on the companies. (This was the 50 percent royalty rate that inspired Saudi Arabia to seek a similar deal.) Perhaps even more worrisome from the companies' point of view, Pérez Alfonzo was working on finding a way to co-operate with the oil producing nations of the Middle East so that the companies would no longer be able to play them off against each other. Those plans—the first glimmer of an organization that would later shake up international politics—were put on hold in November 1948 when a military coup abruptly overthrew the democratically elected Betancourt government. Instead of visiting the Middle East as he had planned, Pérez Alfonzo found himself, along with Betancourt, back in the Puerto Cabello prison.

Thus began another dark era of dictatorship and terror, this time under Colonel Marcos Pérez Jiménez. Although many involved in the former government were killed, Pérez Alfonzo and Betancourt were eventually released and went into exile. Jiménez meanwhile set up a secret police force to suppress dissent, plundered the national coffers and quickly turned himself into the richest man in Venezuela. He also re-established close relations with foreign oil companies, granting them massive new concessions. When he was finally driven out of power ten years later—after widespread rioting, a general strike and a petroleum production halt organized by oil workers—he fled to the U.S., where he was warmly received. (After rebuffing requests for his extradition for four years, Washington finally returned Jiménez to stand trial in Venezuela in 1963. He spent five years in prison there before pressure from Washington secured his release and eventual exile to Madrid.)

With national elections and the return of the AD to power in 1958, Pérez Alfonzo, once again oil minister, resumed his quest for greater national control over the country's oil resources. He declared there would be no new concessions to the companies; the goal would no longer be to produce ever more oil but to maximize the value of existing oil production. Over the years, Pérez Alfonzo had become increasingly interested in conservation and even ecology. Unlike just about everyone else of his generation, he was highly conscious of oil as a non-renewable resource, something to be conserved for the future development of the country, not squandered for short-term bursts of revenue. To manage the resource more effectively, he proceeded with his plan for a national oil company.

Pérez Alfonzo also turned his mind to something even more ambitious: he wanted the oil producing nations to win control over the international oil market. Individually, they were weak and easily pitted against each other by the major companies. Yet the nations had within their borders most of the oil that the world depended upon. So if they acted collectively to limit the amount of crude that made it to world markets, they could keep the price up, conserve this non-renewable resource—and make themselves much richer. What was needed was some mechanism for controlling the amount produced.

That led him, oddly enough, to something called the Texas Railroad Commission, the body charged with regulating oil production in Texas. As we saw earlier, the majors had pushed for such regulation back in the 1930s, when competition from a large number of independents in the U.S. domestic market had threatened to bring oil prices down. Hence the need for some sort of government regulatory body to assign quotas to the various producers in the U.S. market in order to prevent overproduction. This "pro-rationing" was what the Texas Railroad Commission did.

During his many years in exile, Pérez Alfonzo had travelled to the U.S. and studied the operation of the commission in some detail. Now he wanted to take the commission's strategy for managing oil supplies in Texas and adapt it to managing oil supplies worldwide. He made overtures to Washington to find out if it might be interested in working with Venezuela on such a project. After all, the U.S. was an oil producing nation itself. But Washington wasn't at all interested; such a scheme would have no appeal for either American consumers or the major U.S. oil companies. Consumers naturally wanted low oil prices, and any pro-rationing scheme was aimed at maintaining high oil prices, so there would be intense criticism domestically if Washington were to become involved in an international scheme so obviously detrimental to consumer interests. As for the major oil companies, they were already operating their own international pro-rationing system through their cartel, and profiting greatly. There was no advantage to them in sharing some of the decision-making power, and more of the profits, with producing nations like Venezuela. In fact, it was crucial to the majors to be able to play off the producing nations against each other. That way, the companies remained in charge, able to dictate the terms they wanted—which ultimately meant they got away with giving only a small share of the profits to the producing nations.

Washington's rebuff came as no surprise to Pérez Alfonzo. The fact that the Americans weren't interested didn't mean he couldn't find partners for his scheme elsewhere—thousands of miles away, in lands he knew almost nothing about.

From a very young age, Abdullah Tariki was considered to be a bright and capable boy. Even as a youngster in the 1920s, he had proven himself adept at handling the forty-day trek across the Saudi desert on the back of a camel. His proud father had every reason to believe that, when Abdullah grew up, he would take over the family's successful little enterprise running camel caravans across the desert to the big market towns in Kuwait.

However, the political situation in Saudi Arabia changed considerably as Abdullah was growing up. The endless struggles between warring desert tribes gave way to the emergence of the Kingdom of Saudi Arabia. And things changed in Abdullah's life too when, at the age of nine, his cleverness was somehow noted by a courtier travelling with the new king, Abdul-Aziz ibn Saud, in the remote parts of his realm. As it turned out, the young lad never did get to run his dad's camel business. Instead, he would go on to play a key role in reshaping the politics of oil, and in the process shake up the politics of the world. He would become a fervent, uncompromising champion of the rights of the oil producing nations—and a fierce, infuriating opponent of the international oil companies. Howard Page, Exxon's long-time chief Middle East strategist, later described Abdullah Tariki as the most offensive of all the oil nationalists he had had to deal with: "He was the only one I really couldn't stand."

After he was discovered by the king's entourage, it was assumed that Tariki was destined for higher things. Accordingly, he was sent to school in Kuwait for four years and then to Bombay, which was considered an excellent place for up-and-coming Middle Eastern boys to learn the ways of the business world. Upon his return, he received a grant from the Saudi government to go to Cairo, the centre of Arab culture and intellectual life, the very heart of the Arab world. It was the early 1930s, and there was already a growing movement there against the weak, British-controlled Farouk monarchy. Entranced by Egypt, Tariki stayed twelve years and became swept up in the nationalistic renaissance that would later reach its zenith under Nasser.

Tariki returned to Saudi Arabia with lots of radical ideas and a degree in geology and chemistry from the University of Cairo. Another scholarship allowed him to further his studies, this time at the University of Texas, where he acquired both an advanced knowledge of oil geology and an understanding of America, which, needless to say, had almost nothing in common with the nationalistic hotbed of Egypt. Moving to Texas from Egypt was at least as much of a culture shock for Tariki as attending school in Kuwait after a childhood spent camel-driving in the desert. Tariki managed to adapt, staying on in Texas after completing his studies, briefly marrying an American woman and working as a trainee geologist at Texaco.

It was his Egyptian experience, though, that was to have a lasting effect on him. When he returned to Saudi Arabia in 1948, at the age of twenty-nine, Tariki was one of the best-educated men in the country, and the only one with university training in oil geology. He was thus appointed supervisor of oil affairs for the eastern province, the site of the Ghawar oil field. Officials at Aramco were just becoming aware of how truly spectacular Ghawar was—even more so than the rich oil fields of Iran and Iraq. Ghawar was an underground sea of oil 240 kilometres long and 35 kilometres wide—the largest concentration of oil in the world. Pumped at a rate of one million barrels a day (as it was by the mid-1950s), it could have gone on producing for the next 275 years.

Tariki was struck by the wealth that the Ghawar field was generating. He was also struck by how little of it was ending up in the treasury of Saudi Arabia, and how little say his country seemed to have in the management of this magnificent resource. His own "supervisory" capacity had turned out to be largely ceremonial. He found himself totally excluded from Aramco's decision making and largely treated with contempt by company officials, even after he was elevated by the Saudi king to the newly created post of General Director of Oil and Mining Affairs. Increasingly, Tariki felt stirred by the nationalist sentiments that had been awakened in him in Egypt and that continued to reverberate across the Arab world. Much to the alarm of Aramco officials, Tariki began espousing the Nasserite cause in Saudi Arabia, speaking of oil as one of the keys to creating a strong, united Arab world, and even flirting with the idea of the "Arabization" of Aramco—a somewhat vaguer concept than "nationalization," which had landed Iran in so much trouble.

Relations between Tariki and Aramco became even more strained after Tariki pointed out that one of the agreements between the company and Saudi Arabia included a clause, previously unnoticed, allowing the kingdom to appoint two directors to the Aramco board. Tariki got himself appointed by royal decree, and Aramco board meetings were immediately transformed from sedate in-house affairs into volatile confrontations in which Tariki repeatedly demanded access to the company's books. Although Aramco largely refused to co-operate, it did hand over some small bits of information to appease Tariki—and even these small bits were enough to allow him to prove that the company had failed to pay $145 million in taxes. (At the time, the kingdom's annual oil revenues only amounted to $300 million.)

But Tariki wanted to go beyond simply forcing the company to abide by the rules it had agreed to; he wanted to change the rules. And that involved changing the power dynamic. The idea of uniting the Arab oil producing nations had been in play since 1945, with the establishment of the Arab League, which made explicit mention of the Arab world's oil resources in its terms of reference. Actual unity, however, was never more than a far-fetched dream, partly because it was clear that the power of the Arab producers could be easily undercut by a few significant non-Arab producers, such as Iran and Venezuela. In the absence of actual collaboration among the producing nations, the temptation for one country to increase its production at the expense of the others was too great—as had become evident after Iran's nationalization. To compensate for oil they were no longer pumping in Iran, the majors had simply increased their production in other Middle Eastern nations and in Venezuela. All had been happy to get the extra revenue.

Like Pérez Alfonzo in Venezuela, Tariki had come to appreciate the extent to which the companies managed to maintain the upper hand by playing off the oil producing nations against each other. Just as Pérez Alfonzo had thought of reaching out to the Middle East oil producers, Tariki saw the potential of the Middle East making common cause with Venezuela. When he heard about a national oil congress to be held in Venezuela in 1951, Tariki arranged to attend even though he hadn't been invited. With Pérez Alfonzo in exile at the time, there was really no one in Caracas for Tariki to collaborate with. Still, the idea of some sort of co-operation among producers began to percolate in his mind, and on the way home he stopped to visit oil officials in Tehran and Baghdad. He also stopped in Rome, where he met with Enrico Mattei, head of the Italian state oil company ENI, which was provoking the anger of the majors by winning production contracts and market share through aggressive, competitive practices.

The idea of international collaboration among oil producing nations remained a remote dream until conditions changed at the end of the decade. By 1958, Pérez Alfonzo was back running oil affairs in Venezuela and looking for other oil producing nations with which to make common cause. Furthermore, there was a glut of oil on world markets as large new fields were brought into production, notably in an area outside the cartel's control—the Soviet Union. This was creating downward pressure on prices, forcing the majors to sell their petroleum below the "posted" price, which they set themselves. But they were still paying royalties to the producing nations on the basis of the posted price. In February 1959, the majors responded by cutting the posted price by eighteen cents a barrel—a cut that had huge repercussions for countries that relied almost exclusively on oil revenues.

The Arab oil producing nations responded to the crisis two months later by convening the first meeting of the Arab Petroleum Congress in Cairo. Despite the controversy over the price cut, the meeting was fairly low-key; representatives of the companies attended and were not made to feel unwelcome. Significantly, observers from Iran and Venezuela, including Pérez Alfonzo, also attended. So the Cairo congress finally provided a chance for an encounter between Tariki and Pérez Alfonzo.

The two men had heard a lot about each other and, after being introduced, quickly sensed the opportunity to push forward their shared agenda. Ducking out of the formal talks, they headed off to the Cairo suburb of Maadi, where a yacht club, deserted in the off-season, offered a perfect venue for some highly secret talks. They had brought along from the conference a few representatives from the other nations, but it was Tariki and

Pérez Alfonzo who did the talking. With the oil company representatives back at the conference unaware of what was going on, Tariki and Pérez Alfonzo reached a secret gentlemen's agreement at the Cairo yacht club that, as Pérez Alfonzo later said, "constituted the first seed of the creation of OPEC."

A year and a half later, with the oil glut persisting and the market price continuing to sag, Exxon CEO Monroe Rathbone was keen to cut the posted price a second time, a move that everyone knew would be even more explosive than the first cut. The head of Exxon in Venezuela was so convinced that a second cut would be a tactical mistake that he threatened to resign if it was enacted. The Exxon board in New York decided to heed this warning and imposed the price cut exclusively on the Middle East producers—making the cut doubly provocative in countries where Nasser's nationalistic message had already activated a new sense of Arab pride.

A meeting was convened in Baghdad the next month, September 1960, and included five key oil producing nations: Iraq, Iran, Saudi Arabia, Kuwait and Venezuela. Between them they accounted for roughly 80 percent of the world's oil exports, which gave them extraordinary potential power to create havoc in a world dependent on oil. The presence of Iraq, which had not attended the Cairo congress but which had long-standing grievances against the majors, added a militant tone to the gathering. The gentlemen's agreement worked out in Cairo by Tariki and Pérez Alfonzo was now brought into play. Together the two men eagerly presented their plan. The logic was irresistible: if the producing nations agreed amongst themselves to restrain production, they—not the companies—would have control over the world oil market, with the power to set the price. Of course, the companies wouldn't like this new rebelliousness. But Venezuela, which had been spared the second price cut, was proof that standing up to the majors could pay off. Before the Baghdad meeting was over, a new organization was born: the Organization of Petroleum Exporting Countries, a cartel to take on the cartel. A jubilant Pérez Alfonzo declared: "We have formed a very exclusive club . . . we are now united. We are making history."

They weren't quite ready to make significant history, however. The framework of the organization was established, with a head office in Geneva, and there were plenty of meetings and speeches about the importance of confronting the oil company cartel. But for all the heady talk of unity and defiance at the Baghdad summit, the old rivalries and lack of trust quickly reappeared, and no progress was made on the crucial issue of restricting production, which was essential for a pro-rationing scheme like the one in Texas. Then, in March 1962, the new organization received a staggering blow: Tariki lost his job as Saudi oil minister.

Tariki's firing marked the end of a brief flirtation with political reform in Saudi Arabia in the late 1950s and early 1960s. Tariki had been an influential member of a faction trying to push the kingdom towards closer ties with Nasserite Egypt, a nationalistic oil policy and internal democratic political reforms. Led by the dynamic Prince Talal, and enjoying the support of several other Saudi royal princes as well as a small circle of radical intellectuals, the faction had drawn up far-reaching plans to bring Saudi Arabia into the modern world by transforming it from an autocratic monarchy into an open, democratic constitutional monarchy. They envisioned a constitution, a free press and a national assembly in which one-third of the seats would be held by appointees of the royal family and tribal chiefs, while two-thirds would be elected. These moves had been staunchly opposed by Crown Prince Faisal, who served as prime minister. But in December 1960, the increasingly unpopular Faisal was pushed aside by reigning King Saud, who formed a new cabinet in which Talal became finance minister and Tariki took on the oil portfolio. It was a startling moment in Saudi politics. In the West, and particularly

262   WAR, BIG OIL, AND THE FIGHT FOR THE PLANET

inside Washington and Aramco, the prospect of Saudi Arabia embracing Egypt (with its close Soviet ties) and moving towards oil nationalization was viewed with considerable alarm. Talal was dubbed the Red Prince, and Tariki the Red Sheikh.

But the radical experiment was nixed before it got off the ground. Almost immediately, the sickly King Saud announced there would be no new constitution, since his country was ruled by Islamic laws, nor would there be any reforms. By the following summer, Prince Talal's passport had been revoked and his palace raided. A few months later, King Saud reconciled with Faisal, and by 1962, Faisal was once again head of the government and Tariki no longer oil minister. Faisal embarked on a conservative course and, in a move that pleased Aramco and Washington, he replaced Tariki with a charming, urbane, thirty-two-year-old lawyer, Sheikh Zaki Yamani, who loved New York and Western culture as much as Tariki loved Cairo and the Arab world. With Tariki's firing, the effective leadership duo of Tariki and Pérez Alfonzo was now gone from OPEC, and the most significant oil producing nation in the organization (and in the world) was tilting increasingly towards Washington.

Throughout the following decade, OPEC was largely a paper tiger—a forum for loud speeches but little action. The majors and the U.S. government were mostly able to ignore it. James Akins, former U.S. ambassador to Saudi Arabia, recalls that U.S. officials were forbidden even to talk to OPEC, lest such official overtures make the organization seem important.

Meanwhile, Tariki continued to be a thorn in the side of the international oil companies. Banished from the Saudi kingdom in 1962, he travelled extensively throughout the Arab world, constantly giving speeches in which he denounced the oil companies, called for the renegotiation of concessions and urged the establishment of a cadre of Arab technical experts to take over the running of the oil fields. At a packed gathering of the Arab Oil

Congress in Beirut in November 1963, Tariki gave a rousing address that became legendary in the Arab world. "When God endowed Iran, Venezuela and the Arab countries with vast reserves of oil, He did not do that to give the chance to foreign people to come from far countries to exploit those reserves," Tariki exclaimed, the audience enthralled. He argued that foreign companies were no longer needed to develop Middle Eastern oil, and warned that failure to take control of the oil would have disastrous consequences for the Arab world. "The Arab nation is living in a period when extraordinary means of development are available to it. This chance may not be repeated. If the Arab nation does not make good use of it, it will be lost and we shall have condemned our coming generations to everlasting poverty and underdevelopment."

Tariki settled in Lebanon and, along with Lebanese oil expert Nicolas Sarkis, established an oil journal that quickly became a focal point for the nationalistic intelligentsia of the Arab world. He later started up his own journal called *The Oil of the Arabs* (with the subtitle *Arab Oil for the Arabs)*. He also served as an oil adviser to the more nationalistic regimes in Iraq and Algeria, and later to Qaddafi in Libya. Indeed, it was as an adviser to Qaddafi that Tariki finally delivered a serious blow to the international oil order. By coming up with the strategy of using the independents as leverage against the majors, Tariki outsmarted Exxon and forced the company to accept Libya's 40-cent-a-barrel royalty hike, which began a spiral that spread through the Middle East. It's not surprising that Exxon's Howard Page concluded that Tariki was the "one I really couldn't stand."

In Venezuela, Pérez Alfonzo soon became disillusioned with politics, as the government he was part of moved increasingly to the right, rejecting the nationalistic oil policies that he had pushed so hard for. "At the end, he was excluded from all government decision making," says Hugo Chávez. Retreating into private life,

Pérez Alfonzo devoted himself to his ideals of conservation and ecology, and only occasionally surfaced to dismiss OPEC as "too passive." Passivity was not something that ever sat well with Pérez Alfonzo, who once had been overheard angrily telling oil company officials, "What you give us is the crumbs and bones of your feast; but we are not dogs."

OPEC would never live up to Pérez Alfonzo's vision of an instrument of empowerment and a vehicle for conservation, but it would soon at least ensure that its members were admitted to the feast.

———

By the fall of 1973, things had changed dramatically in the world oil situation. The glut that had plagued the industry, bringing down prices so significantly in the 1950s and 1960s, had disappeared. If anything, oil supplies were tight, as demand was growing faster than new sources of supply. Traditionally, when a world oil shortage developed, the U.S. was able to bump up its own production to make up for the shortfall. But after fourteen years of import controls, the U.S. had been using up its own reserves at an astonishing rate, leaving it with little capacity to quickly bump up production levels in a crunch. All this helped enhance the power of those who had oil to sell.

The bargaining power of the oil producing nations had already been rising, following Libya's success in winning a better deal from the industry in the late 1960s. That had led to rising expectations and demands in other producing nations. The idea that nations should own their own oil industries had also become widely accepted throughout the Middle East, with the debate raging over whether there should be full nationalization (as favoured by Tariki), or simply "participation" as a step towards full nationalization (as favoured even by moderates like Yamani). In 1971 both Libya and Algeria had nationalized significant parts of their oil industries, and in 1972 Iraq finally nationalized IPC. The small Gulf states of Abu Dhabi, Kuwait and Qatar negotiated participation deals giving them 20 percent ownership. And by the end of 1972, even pro-Western Saudi Arabia had insisted on getting a stake in the majors' biggest prize—Aramco. Yamani had brokered a deal in which the companies agreed to give the Saudis 25 percent of Aramco, rising to 51 percent after ten years.

The oil producing nations had also started to assert their power collectively in the early 1970s. Indeed, the West had begun to see OPEC as a potentially formidable force. After Libya's success prompted similar demands elsewhere, the majors had actually decided it was in their interest to negotiate collectively with the producing nations rather than allow producer demands to "leapfrog" from one nation to another, with ever higher expectations. In showdowns with the producing nations in Tehran and Tripoli in 1971, the industry had agreed to a price increase—an increase that was supposed to hold for five years. But in October 1973, OPEC called the majors to a meeting in Vienna with the unabashed purpose of scrapping that deal and winning a significantly higher price. The days when American officials could blithely refuse to talk to OPEC were gone.

As both sides assembled for the Vienna confrontation, the whole situation suddenly became a lot more volatile. On October 6, Egypt and Syria invaded territory occupied by Israel, and Arab anger over ongoing U.S. support for Israel made the oil price negotiations, now moved to Kuwait, all the more tense and bitter. The industry's offer of a 15 percent price increase was rejected outright by the producing nations, which countered with a demand that prices be doubled. Shocked by the militancy and apparent unity of OPEC, the oil companies decided to consult with the major western governments. The answer came back clearly: Don't give in. With negotiations stalemated, on October 16 OPEC did something it had never done before: it raised the price of oil itself—by a staggering 70 percent. The cost of barrel of a barrel of oil shot from $3 to $5.11. Up until this point it had always been the companies that had set the price, declining even to consult with the producing nations. Now the nations had set the price—without consulting the companies.

The next day, the Arab members of OPEC met by themselves (in an organization called OAPEC) and upped the ante further, releasing a communiqué calling for an immediate cut in oil production of 5 percent and vowing to reduce production by the same amount each month "until the Israeli withdrawal is completed from the whole Arab territories occupied in June 1967 and the legal rights of the Palestinian people restored." When the Nixon administration three days later announced a $2.2-billion military aid package to Israel, the Saudis—the most pro-American of the Arab countries—declared they were cutting off all shipments of oil to the U.S.

The use of oil as a weapon had long been contemplated and threatened, and was even tried unsuccessfully during the 1967 Arab-Israeli war. Back then its effectiveness had been blunted by the ability of the U.S. to push up its own production. This time, however, with the U.S. unable to compensate for the shortfall from its own domestic reserves, the embargo was far more effective. The gradually escalating cutbacks announced by the Arab producers, although aimed at increasing leverage against Israel, also contributed to the scarcity of oil, and thereby helped reinforce OPEC's dramatic price hike. This was exactly the kind of strategy that Pérez Alfonzo had envisioned (although not connected with the Israeli situation): sell less oil, but sell it at a higher price, thereby realizing the same or higher revenues while preserving oil reserves for the future.

And the price jolt wasn't over yet. Less than three months later, OPEC met again in Tehran and jacked up the price once more, this time more than doubling it—from $5.11 to $11.65.

(Particularly staggering was the fact that this latest hike was actively pushed by the shah, despite his long-time closeness with Washington.) In four years the price of oil had jumped from its long-established level of $1.80 a barrel to $11.65, an increase of more than 600 percent.

Between the price hikes and the embargo, the world suddenly found itself experiencing an energy crunch. The Arab world had been producing 20 million barrels of oil a day before the embargo, and two months later—at the embargo's height—that amount had declined to 15 million barrels. While the effects were felt just about everywhere, they were particularly acute in the U.S., which, because of its stalwart support for Israel, was the main target of Arab wrath. Within a few months gasoline prices in the U.S. had shot up 40 percent. If that wasn't jolting enough for a nation whose culture revolved around the car, there was the new phenomenon of gas station lineups, with gas purchases restricted by the days of the week or by whether one's licence plate had a final number that was odd or even. Things like that weren't supposed to happen in America.

Dramatic as it was, the crunch didn't last very long. Within a few months its effectiveness was blunted as more and more Middle Eastern oil somehow evaded the embargo and found its way back onto the market. Meanwhile, the Arab–Isaeli war had ended and Egyptian leader Anwar Sadat, who had been a keen advocate of the embargo when his country had launched the war, had been wooed by Washington into becoming an American ally, and was advocating an end to the high-pressure oil tactic. In March 1974, the Arab oil ministers agreed to end the embargo, convinced partly by arguments that such a move would pave the way for a serious U.S. effort to advance a solution to the Middle East conflict. In all, the embargo had lasted five months.

Still, the embargo and the ongoing high oil prices were to have a deep, long-lasting effect on the world of oil, and on politics in general. For the first time, an upstart group of nations—mediocre players on the international scene—had banded together and out-manoeuvred the West, leaving the most powerful nation on earth temporarily unable to fully indulge in the kind of extravagant, free-wheeling lifestyle that millions of Americans had come to consider their birthright. Clearly, this kind of audacity would have to be curbed; the unquestioned dominance of the West would have to be restored.

———

The first thing to note about the impact of the 1973–74 energy crisis and its aftermath was how favourable it was for the oil companies. For all the struggling that went on in the lead-up to the explosive price increases of 1973—with the oil companies facing off against the oil producing nations in countless sessions in Tehran, Tripoli, Geneva and Kuwait—the simple truth is that the major oil companies emerged huge winners, enjoying profits beyond anything they'd seen in the previous thirty years. In 1973, the year OPEC unilaterally imposed a fourfold increase in oil prices on the world, Exxon reported a profit of $2.5 billion, which at that point was an all-time record for any corporation on earth. The following year, with the West still struggling to readjust to the shock of the new energy era, Exxon's profits soared again. One can best appreciate the magnitude of Exxon's gain by tracking its rate of return over a period of years. From 1963 to 1972 the company had enjoyed a comfortable overall return averaging 12.8 percent. In 1973 that return surged above 16 percent, and in 1974 it reached an astonishing 21.3 percent. While Exxon led the pack, the other majors also did exceptionally well in 1974, with Mobil reporting a profit of 17.2 percent, Standard (Indiana) 21 percent, and Gulf 17.9 percent, for an overall average among the majors of 19 percent.

This raises the intriguing question of what role the companies may have played in contributing to their own good fortune. Daniel Yergin seems to attribute their windfall largely to good luck. "Much of the immediate increase derived from foreign operations," he writes. "As the exporting countries pushed prices up, the companies got a free ride." But were the companies really just passive observers, watching in amazement as money kept flowing in, while OPEC pulled all the strings? John Blair, the antitrust economist, questions the notion that the companies were largely passive beneficiaries, noting that, at the time, OPEC lacked the technical expertise required to carry out the enormously complicated task of adjusting oil supplies worldwide to maintain the desired price level. "If output was to be effectively controlled by the OPEC countries, they would have to develop some mechanism of control similar to that perfected over a long period of years by the companies," Blair observed.

The notion of OPEC deftly pulling the strings—popularized in the mainstream Western press—was never taken seriously in the oil press, where the complexity of oil pricing is appreciated. The *Petroleum Economist* noted early in 1975, for instance, that while the producing nations had long considered the possibility of "programming" their production, this was not easily accomplished. "[I]ts effective implementation necessitates fixing basic quotas for individual producing countries and devising machinery to enforce production decisions and impose sanctions against non-compliance. At present, this loosely-knit organization [OPEC] simply does not have the staff or the expertise to operate as an effective cartel."

If OPEC lacked the sophistication to operate as a serious cartel, the majors certainly did not. This was precisely what they had long excelled at, dating back to the Achnacarry accord of 1928. Blair concludes that, with OPEC's lack of competence and authority in such matters, "the function of precisely tailoring supply to a diminishing demand must have been the work of the companies which . . . simply continued their historic role of stabilizing the market at the existing level of price."

The role played by the companies in keeping supplies tight during the energy crunch was implicitly acknowledged by Exxon chairman Clifton C. Garvin in a televised interview on *Face the Nation* in November 1975. Morton Mintz, a *Washington Post* reporter, pointed out on the program that Aramco had cut back production in Saudi Arabia by 2.4 million barrels a day the previous month. "Now doesn't this prove that Aramco, and not Saudi Arabia, is calling the shots and doesn't this prove that without the support of Exxon and the other major oil companies, the cartel [OPEC] would collapse and the prices of oil would drop?" Garvin immediately denied all this. But he then proceeded essentially to confirm what Mintz had just suggested—that it was the company, not Saudi Arabia, which had made the decision to cut oil production. He explained why they did so: "Europe is having a big recession; Europe is having a mild fall so far. There just hasn't been the demand. As a result, *we've had to cut our production of oil.*" Later, when the question came up again, Garvin elaborated on his earlier answer, noting "*our estimates* of what the oil demand was going to be just didn't materialize, so *we had to cut back* [italics added]." The Exxon chairman was admitting that the company's recognition of slumping oil demand led it to cut production and it was this production cut—described at the time by *Petroleum Intelligence Weekly* as "stunning"—that allowed oil prices to remain high.

Economists and government authorities had been predicting a price collapse as a result of the 1975 recession, and the decline of OPEC's power. The slump, which had been caused by higher oil prices, meant there would inevitably be lower demand for oil. Thus, the reasoning went, there would be an oil glut and the price would come tumbling down—to the great relief of Western consumers. But no such relief happened because the majors spotted the "problem" with their sophisticated monitoring and moved to prevent an oil glut from occurring, by making a big cutback in production. In other words, the majors had played a crucial role in allowing OPEC to maintain its high prices. The oil companies knew how to make the system work, and they managed it effectively for their own benefit (and, incidentally, for OPEC's).

As Western rage flared against the Arab oil producers—a rage that would simmer over the years and provide a strong back wind for many U.S. policies—the oil companies were content to look like innocent free riders, with their multi-billion-dollar profits the result of nothing more than the good luck of happening to be in the right place at the right time.

The rise of OPEC was not easily digested by the West. The notion that a ragtag assortment of lightweight countries— Venezuela, Saudi Arabia, Iran, Iraq, Kuwait . . . Libya, for God's sake!—could be dictating policy in the world economy was not something that seemed justified, at least not to those accustomed to dictating policy themselves. So, in the West, this turn of events was presented as some sort of violation, an aggressive assault on the state of affairs that had long existed among the world's law-abiding nations. U.S. president Gerald Ford declared in September 1974 that "[s]overeign nations cannot allow their policies to be dictated, or their fate decided, by artificial rigging and distortion of world commodity prices."

As we've seen, artificial rigging and distortion of world commodity prices had been going on in international oil markets for decades, but the rigging and distortion had in the past been carried out by the major oil companies—an arrangement that sovereign nations like the U.S. not only allowed but actively encouraged and, at times, enforced with military power. The key decisions about the international oil market had long been made in boardrooms in London and New York, not to mention castles in Scotland. These decisions—which had enormous repercussions on the oil producing nations, had nothing to do with the operation of a free market. On the contrary, they were the very antithesis of a free market. Instead of open competition, the international oil market had been carefully divvied up among a small group of players actively colluding with each other. Furthermore, the decisions made by this clique were arbitrary, not governed in any way by free market principles. Iraq's production was held down while Saudi Arabia's expanded, for reasons that had nothing to do with any differences in the marketability of Iraqi and Saudi oil and everything to do with punishing defiant countries and rewarding co-operative ones.

And in Iran, the Mossadegh government discovered just how arbitrary the rules of the international oil world were when it nationalized Iranian oil. Washington initially sent two high-level officials, diplomat Averell Harriman and oil adviser Walter Levy, to Tehran to explain to Mossadegh the realities of the oil market. Mossadegh told them he planned to sell Iranian oil for whatever the market would bear. The two U.S. officials grew exasperated trying to communicate to him a simple fact: there was no free market for oil! The majors set the price. That was the way it worked. When Mossadegh went ahead with his nationalization plans, the oil industry, as we've seen, responded with a boycott, effectively closing markets all over the world to Iranian oil. This boycott was a brazen assault on the operation of a free market— yet the British and U.S. governments didn't intervene to defend free market principles. On the contrary, they overthrew the Iranian government.

The Western condemnation of OPEC for artificial rigging and distortion of the oil market is therefore stunning. As long as the oil companies were running the show, rigging and distortion were fine, there had been no denunciations from the West, everything had been considered to be functioning properly. It was only when the oil producing nations themselves started to engage in rigging and distorting that the situation was seen as out of control, as an affront to the sovereignty of nations.

Of course, the real concern was that money was ending up in the wrong part of the world. The revenues of the producing nations rose from $22 billion in 1973 to a staggering $90 billion the following year—a shift that had huge implications throughout the world. "Rarely, perhaps never before, had a group of countries managed to increase their income so suddenly," notes French oil analyst and historian Pierre Terzian. The media focused on images of rich Arab oil sheiks, who instantly became symbols of greed and anti-Western treachery. Certainly, there were many overnight multimillionaires, particularly those with royal connections in the sparsely populated Gulf states of Saudi Arabia, Kuwait, the United Arab Emirates and Qatar.

These four countries were hardly typical of OPEC nations, which together accounted for some 300 million people, the majority of whom lived in fairly primitive conditions. Average income in those four Gulf states more than doubled, from $3,528 in 1973 to $7,600 in 1974. But in OPEC countries with larger populations and lower oil production, gains were much smaller, with average annual incomes rising from a modest level of $396 to a still-modest $540. (By way of comparison, it's worth noting that the average income in the industrialized countries was $5,203 in 1974—1 percent lower than it had been in 1973.) Meanwhile, in Nigeria, an OPEC country with 80 million people, the increase in oil revenues raised the average income by just $80 per person. And in Indonesia, with a population of 130 million, the average income increased by a meagre $18. The rich oil sheik, colourful as he was for media purposes, was hardly a ubiquitous character outside of a sprinkling of palaces in the Gulf.

But the notion that there was a plot by Arabs aiming to take over the world pervaded the media. *The Economist* informed its readers that it would take only 15.6 years for OPEC to use its surplus to buy up every publicly traded company in the world. The Parisian journal *L'Express* made a similar sort of calculation to show that, with its spare cash, OPEC could take over the Champs-Elysées in ten days, and in just eight minutes it could buy up the Eiffel Tower. Interestingly, it was never pointed out how quickly Exxon could buy up the Eiffel Tower or the Champs-Elysées. Of course, the French government was unlikely to put such national icons up for sale in any case, so the concept was a red herring—but a useful one for stirring up animosity towards OPEC and generating fear that the Arab world would use its new-found wealth to tear the heart and soul out of the West.

There was constant talk in the media of how the enormous sums of oil money could be "recycled," the implication being that there was something inherently wrong with so much wealth finding its way over to those faraway parts of the world when the proper place for vast quantities of money was here, where *we* could use it. In fact, much of the OPEC surplus *was* "recycled" back into the economies of industrialized countries. One analysis by the Bank of Chicago found that of the total gross foreign assets of $160 billion accumulated by OPEC by 1977, about $42 billion was invested directly back in the U.S., $16 billion in other industrialized countries and some $60 billion in funds controlled by American banks.

In all the ominous talk, the underlying suggestion was that the distribution of world income prior to the emergence of OPEC had been natural and just. The fact that there were huge concentrations of wealth in the hands of a relative few had not been considered a problem in the West—until some of that highly concentrated wealth ended up in the oil producing world. Interestingly, in an earlier period, in the 1950s and 1960s, there had been a different sort of shift in world income patterns, when a large drop in the price of raw materials (including oil) had lowered the income of many poor nations. At the same time, industrialized countries had benefited from the rising price of manufactured goods. "This produced huge trade deficits for the developing countries, while it was regarded by the developed countries as a normal market situation of supply and demand," noted Ali Attiga, the Libyan economist, who also served as secretary-general of OAPEC. It was only when the shoe was briefly on the other foot, with the income shift working against the West, that commentators here even noticed that a shift had taken place.

Furthermore, one can make a case—as the oil producing nations have—that for years their oil was sold at bargain-basement prices, and that this greatly helped finance the rapid industrial growth and prosperity of the West. It is certainly true that oil is a crucial raw material in the industrial age. Given its relative scarcity, non-renewable nature and high efficiency as a source of energy, it is an unusually valuable item by any standard, and its price should logically reflect this value.

The underpricing of oil seems particularly unfair because most of the countries that have large quantities of it have virtually nothing else in the way of resources. This was Abdullah Tariki's point in his powerful address to the Arab Oil Congress in Beirut in 1963. For many oil-rich nations, this is their one chance. Once their oil is gone, it is never to return, and even countries with the largest supplies are expected to see those depleted by the middle of this century. So the oil-rich nations have a brief, one-time window of opportunity to make use of their oil in a way that will position their countries well for the future. To underprice such an asset would be a reckless and irresponsible course that would gravely damage the interests of future generations in those countries. Within a few decades—certainly within the next fifty years—they will be without their main source of income. Imagine if the West faced such a prospect; tanks would already be mobilizing.

Seen in this context, OPEC starts to seem like a rather understandable organization. Even though it effectively operates as a cartel, adopting some of the same anti-competitive practices that the companies' cartel did, OPEC is doing so to enrich what are, in most cases, poor countries with few other resources. Moreover, the oil is at least *their* oil. The companies, on the other hand, managed to corner the oil market by gaining control over reserves in faraway lands, often through trickery, bribery and intimidation, and sometimes even with military assistance.

Of course, the 1973–74 embargo did seem to introduce a whole new level of aggression into the artificial rigging of markets. All of a sudden a vital commodity was being physically withheld, raising the possibility of serious deprivation, of the West being "held hostage." Dramatic as the concept is, however, it is hardly new, nor is it confined to this situation. For decades Washington has imposed a tight embargo on Cuba, preventing many vital commodities from entering that country and creating great hardship there. Similarly, the 1950s boycott against Iran crippled that nation's economy. More recently, throughout the 1990s, the UN Security Council, at the insistence of Washington, maintained devastating sanctions on Iraq—sanctions that deprived Iraq of medicine and food, and that have been linked, by former U.S. attorney general Ramsey Clark, to an estimated 1.5 million Iraqi deaths.

The response of many in Western media and governments presumably would be that these countries deserved such harsh treatment for their bad behaviour (though it's hard to imagine what kind of system of justice allows 1.5 million people to die because their unelected dictator behaved badly). By comparison, the West holds itself blameless in connection with the 1973 oil embargo, so the fact that Americans had to line up for gasoline seems like an intolerable infringement on their rights. The morality of the situation might well seem very different to Arabs living in the Middle East. In launching the war in October 1973, Egypt and Syria were trying to reclaim some of their own territory, which Israel had taken from them in the 1967 war and which repeated UN resolutions had called for Israel to relinquish. If one accepts the notion that nations have a right to protect their territorial integrity—as we in the West certainly do when it comes to defending our own borders—then it shouldn't be surprising that Egypt and Syria would feel entitled to try to regain their lost territory, and that fellow Arab nations would support them by withholding oil as a pressure tactic. Indeed, as pressure tactics go, the oil embargo was pretty mild; gas lineups may have been annoying, but nobody died in them. The Arab oil embargo seems like an outrageous abuse only if one believes that embargoes—which the West readily uses to punish "rogue" behaviour around the world—should be a weapon only in *our* arsenal.

———

The ultimate argument used by the West to vilify OPEC was that the oil price hikes were devastating to poor nations. It certainly was true that the higher oil prices had a disastrous impact on most Third World nations, which were least able to afford the higher prices and most desperately in need of energy to advance their economic growth. In 1974, the developing countries faced an added $11 billion in oil costs, causing their collective trade deficit to shoot up to $35 billion, four times what it had been in 1973. There were pleas from developing countries for relief; a group of African nations requested a two-tier pricing system that would allow poor nations to be spared from price hikes. OPEC considered the request at meetings in January and June of 1974, and both times rejected the idea. So the West's frequent denunciations of OPEC for its indifference to the plight of the Third World seem to have some merit.

A closer look suggests that the picture is not so clear. First, it should be noted that there was something deeply ironic about prominent Western figures suddenly taking such a keen interest in the plight of Third World countries. Henry Kissinger, of all people, rarely missed an opportunity to mention the impact of the oil price hikes on Third World nations, which, he noted, "face the threat of famine and the tragedy of their disappointed hopes for future development." Of course, Kissinger had never been particularly concerned about these countries in the past, in the days when "their disappointed hopes for future development" couldn't be laid so squarely at the feet of his arch-enemy, OPEC.

Also noteworthy was the industrialized world's limited willingness to help the countries whose fate it so keenly lamented. Despite promises to set aside 0.7 percent of their total gross national product (GNP) for foreign aid, the industrialized Western nations collectively mustered barely 0.3 percent throughout the 1970s. A few impressive exceptions were the Netherlands and Norway (both above 0.6 percent) and Sweden, which reached 1 percent. Meanwhile, Germany contributed only 0.4 percent, and the U.S.—the world's richest country—a meagre 0.2 percent.

These numbers provide an interesting point of comparison with the fairly significant financial help given to the Third World by OPEC countries in the wake of their oil price hikes. Even before the hikes, OPEC countries had a record of giving away a higher level of their GNP in foreign aid. As OECD statistics show, OPEC countries allocated considerably higher levels to foreign aid: 1.2 percent of their collective GNP in 1973. After the price hikes, they boosted foreign aid to 2 percent of GNP in 1974 and 2.7 percent in 1975. Notes Terzian: "These efforts were considerable from any point of view"—especially since many of the big, populous OPEC countries were far from rich. It's also striking that the Gulf sheikdoms, which had become practically synonymous with greed in the Western mind, gave foreign aid at rates that would have taken our breath away, had we ever heard about it in the West. According to the OECD, Saudi Arabia allocated nearly

5 percent of its GNP to foreign aid, while Kuwait, the United Arab Emirates and Qatar gave away *10 to 15 percent* of their GNPs. One could say that these Gulf states certainly could afford to do so, but then the same could be said for us in the West, and our track record is not nearly so good.

Even more compelling in its vision and ambition was the bold and impressive attempt on the part of a small group within OPEC, led by Algeria, to seize the moment and use the rise of OPEC as a means of bringing about far-reaching changes in the world economy. Starting with a special session at the United Nations in New York in April 1974, Algeria called for an action program aimed at scrapping the system under which the world economy was ruled by the tiny elite of G7 nations, and opening it up to participation from a much broader group, dubbed the Group of 77. The UN General Assembly adopted the plan, which called for nothing less than a "new world economic order."

Easier said than done, of course. But enormous effort was put into furthering the plan, led by Algerian president Houari Boumedienne. The central idea was to prevent the West's response to the new realities created by OPEC from focusing exclusively on the issue of oil prices. Boumedienne insisted that any alteration of oil prices would have to be part of a larger re-examination of how the world economy worked, how unfairly it treated producer nations and how badly Third World nations were exploited by the industrialized West.

Boumedienne realized that the oil price hikes had opened up an opportunity that had never existed before. "In the past we spoke and nobody listened. We shouted and heard no echo. We called for help and the wind drowned our voices," the Algerian leader said in early 1975. Now the power of oil had delivered to OPEC the rapt attention of the West. The organization that the U.S. had refused to speak to only a couple of years earlier was being actively courted—as well as vilified—by Western institutions anxious to engage it in "dialogue." Boumedienne wanted to head off the possibility that the OPEC nations would be cajoled into simply joining the club of rich Western nations, with special status as crude oil producers. "We must be careful not to fall into the trap which has been laid for us. We should refuse to talk with the industrialized countries simply in our capacity as oil producers." He stressed that there would be no dialogue on the narrow terms the West wanted. There would be no isolated fixing of the "oil problem." The entire problem of the raging inequality between the North and the South had to be on the table.

It was a heady proposition. As Hugo Chávez is attempting today, Boumedienne was carving out a role for OPEC as champion of the Third World, as a group of Third World nations that would use their new-found wealth and power not just to secure for themselves a privileged niche in the global economy but to bring the rest of the Third World along with them. "If we have to freeze prices, we shall freeze them. If we have to lower them, we shall do so," Boumedienne declared. "On the condition, however, that there is an equal and simultaneous effort on the part of the developed countries, each contributing according to his means and responsibilities to the reordering of the world economy."

With Algeria as the catalyst, OPEC proceeded to develop its plan at meetings throughout 1974 and at a congress that brought together OPEC and other Third World countries in Dakar, Senegal, in February 1975. The following month nearly a thousand delegates flocked to a summit in Algiers, a city awash in banners proclaiming "OPEC is the shield of the Third World." The summit was in many ways a breakthrough, with OPEC members going far beyond their traditional focus on prices and taxation issues, and endorsing a "solemn declaration" stating that "the world economic crisis . . . stems largely from the profound inequalities in economic and social progress between peoples . . .

and that these inequalities . . . have been engendered and maintained mainly by foreign exploitation." Issuing from countries that, only a few years earlier, had barely dared to question their foreign masters, these were bold words indeed.

But in the end, the summit rejected most of the ambitious agenda that Algeria had laid out. Most OPEC nations were pleased to espouse Third World solidarity but were also keen to make peace with and establish closer connections to the West. They rejected Boumedienne's plan for a "development fund" with capital of $10 to $15 billion, preferring to leave foreign assistance to a more ad hoc system in which countries would individually determine their level of aid. The OPEC members also rejected the idea of a joint defense pact.

It goes without saying that the West was largely cool to the idea of rethinking the whole world economic order. (Instead, Western nations established the International Energy Agency to co-ordinate emergency oil sharing and stockpiling programs, so they'd be less vulnerable in the future.) Among Western countries, France was most responsive to the OPEC initiative, agreeing in principle to a thorough re-examination of the rules of the global economy. At the behest of the French, an international "dialogue," not restricted to oil issues, was begun in Paris. Called the International Conference on Economic Co-operation but dubbed the North–South Dialogue, it included all the major industrialized countries plus seven from OPEC and twelve from the Third World. For the following two years the "dialogue" dragged on in endless meetings, co-chaired by Canada's external affairs minister, Allan MacEachen, who had a long-time interest in social justice issues, and Venezuela's Perez Guerrero, a former oil minister and close associate of Pérez Alfonzo. But despite the dedication of the co-chairs to the cause of economic development, they were unable to move the industrialized countries to address any of the fundamental problems raised by OPEC and the other

Third World nations. The rich nations would agree only to boost their aid to 0.7 percent of GNP—a target they failed to achieve—and to create a fund to shore up the price of raw materials from developing countries. In the end, no such fund was ever created.

————

Sadly, the reformist zeal also waned within OPEC. A significant turning point came in March 1975, only three weeks after the Algiers summit, when King Faisal was murdered by an apparently deranged member of the Saudi royal family. Although a conservative in many ways, Faisal had respected Boumedienne and worked with him to achieve consensus within OPEC, thereby creating a link between Boumedienne's radical faction (Algeria, Libya and Iraq) and the more conservative Gulf states led by Saudi Arabia. Faisal was also a committed nationalist who would, as Terzian notes, "on occasion say no to the United States, despite his very strong and extensive links with that country." To some in the U.S., even this moderate degree of independence had been unbearable. Faisal was once described, astonishingly, in *The Washington Post* as having done "more damage to the West than anyone since Hitler."

Faisal's death changed Saudi politics dramatically, and in ways with far-reaching consequences for both OPEC and the world. The Saudi crown passed to Prince Khalid, who was widely regarded as a weak and ineffective monarch, while a power struggle took place between two leading princes—Fahd and Abdullah, both brothers of Faisal but with different mothers, who represented rival clans. Significantly, the two had different approaches to the key issue of relations with the U.S. While both princes considered there was a need for an alliance with Washington, Abdullah was less Western-oriented and had more of Faisal's independent streak. Unlike just about everyone else in the top levels of the royal family, Abdullah had actually chosen to maintain a desert life, living in tents, sipping camel's milk and rejecting the palaces of Riyadh

and the yachts of the Mediterranean. For Abdullah, it was important to maintain the Islamic and Arab character of Saudi Arabia,
in both domestic and foreign policy, and he worried that too much
oil production would bring modernization and corrupting influences. Fahd, on the other hand, had strong links to Washington
and was keen to develop those links further. It was Fahd who
became crown prince after Faisal's death, and Fahd who directed
the nation's oil policy.

The ascendancy of Fahd, who became king in 1982 upon
Khalid's death, gave new life to the "special relationship"
between Riyadh and Washington. Increasingly, it meant that, in
exchange for guaranteeing the security of the royal family,
Washington would be permitted to determine the amount of oil
Saudi Arabia produced. With roughly one-quarter of the earth's
oil, Saudi Arabia could effectively undermine OPEC's high prices
by flooding the market with extra quantities of Saudi oil. This
was an enormous power, which only the Saudis possessed, and it
was exercised more and more often at the behest of Washington.

So, for instance, when a majority of OPEC members pushed
for a 20 percent price hike (to compensate for inflation) at an
OPEC meeting in Bali in May 1976, Saudi Arabia was firmly
opposed. This led to a fierce dispute between Iraq and the Saudis.
The dispute dragged on for months and eventually prompted
Sheikh Yamani, the Saudi oil minister, to threaten publicly to
flood the market with oil and bring the OPEC price tumbling
down. Yamani was denounced throughout the Arab world as a
traitor, while being hailed as a moderate in the Western press.
Indeed, Yamani—and Crown Prince Fahd, who was calling the
shots—often seemed to favour U.S. interests over those of the
Arab world. When the Saudis again resisted a major price hike in
1978, holding the increase to 10 percent, Yamani told Western
journalists, "I did what I could to contain the increase as much as
possible." Lebanese oil expert Nicolas Sarkis denounced Yamani,

reminding him that "OPEC was not set up to be a subsidiary of the U.S. State Department, but to defend the interests of a group of underdeveloped countries."

With the close relationship between Washington and Riyadh solidifying in the late 1970s, the days of OPEC as a cohesive force defending the interests of its members—let alone the broader interests of the Third World—were truly gone. Saudi Arabia was producing a massive 8.5 million barrels of oil a day, creating enough supply to keep oil prices from rising, so that, over time, they were significantly eroded by inflation. Riyadh's high-volume production—unnecessary for its own purposes but favoured by the U.S. in the aftermath of the energy crisis—also allowed the West and the Western oil companies to build up their own oil stockpiles. With large stocks, the companies were able once again to play a role in setting oil prices, by buying and selling large quantities of oil for immediate delivery on the so-called "spot market."

When a second round of oil price shocks occurred after the 1979 Iranian revolution, it wasn't because of actions by OPEC. Rather, the price hikes resulted when the oil companies bid up the price in spot market transactions, pocketing billions of dollars in the process. This prompted an angry response from even the Saudis, who called on Washington to intervene to stop the oil companies from "increasing the price of oil and selling it at higher prices than those set by OPEC." But the Saudis were largely responsible for their own predicament—and for the predicament of all members of OPEC. By flooding the world with oil to please Washington, the Saudis had undercut OPEC's ability to maintain a high price. This had pitted OPEC members against each other, as they scrambled for market share amid falling oil prices. With OPEC badly divided, the dream of an organization with the clout to challenge Western corporate power had been destroyed. Control over oil policy, which had been boldly seized for a brief moment by a small group of underdeveloped nations, had effectively been reclaimed by the West and the international oil companies.

———

Ironically, the oil price hikes and embargo of 1973–74, routinely dubbed an "energy crisis" in the West, supplied a badly needed jolt that at least slowed down the pace with which we were recklessly churning carbon dioxide into the air. Until gasoline briefly became scarce and expensive, almost no attention had been given to the goal of using less of it. Had we continued on that wildly irresponsible course—instead of switching to the slightly less wildly irresponsible course we're now on—we would presumably be that much farther down the road towards the full onset of whatever global warming has in store for us. So, for all our vilification of OPEC, its moment of militancy may at least have bought us a few more years of relative equilibrium on the climate front.

Furthermore, it turned out that cutting back our use of oil wasn't nearly as difficult as one might have assumed. As Yergin points out, by 1985 the U.S. was 25 percent more energy-efficient and 32 percent more oil-efficient than it had been in 1973. Most of the savings hadn't come from people going without energy but rather from technological improvements—notably in car engines and appliances—which simply enabled our machines to perform the same functions but more efficiently. In any sane assessment this amounted to a net improvement, even a financial saving. Ralph Torrie, an Ottawa energy consultant, notes that the energy saved through increased energy efficiency over the last three decades is actually the biggest new "source" of energy we have. And we've gained this new source, he notes, "[w]ith almost no government assistance, in the absence of well-organized institutional and financial infrastructure for its delivery, and against heavily subsidized and highly organized competition from oil, gas and nuclear power." Asks Torrie: "How much more could we get from this source if we *tried*?"

Thus, in many ways, the real "crisis" of 1973–74 wasn't the sharp rise in energy prices but our failure to regard these higher prices as a wake-up call alerting us to the need to cut back our oil consumption. Despite the ease with which that consumption could have been reduced, we have largely refused to reduce it. The panic over oil supplies disappeared almost as quickly as it had appeared, and with the panic went much of the resolve to change our wasteful energy ways. By August 1974, only a few months after the oil embargo was lifted, a bill to spend $20 billion on badly neglected mass transit systems in American cities was cut almost in half. The supremacy of the car, it turned out, was not going to be challenged.

Indeed, nothing would really be challenged; nothing would really be changed. There would be no meaningful attempt to reduce oil consumption, nor would there be any serious effort to address the global inequalities that OPEC, with the oil weapon in its back pocket, had briefly managed to bring to centre stage. The message was clear: back to your tents and your barrios, boys, the West is still in charge.

Having rejected solutions that could have benefited the environment and made inroads against world poverty, the United States instead focused on whittling down the power of OPEC. And increasingly, Washington set its sights on an even more ambitious, long-term solution to the energy crisis: taking physical control of the oil fields of the Middle East.

CHAPTER 9

KING OF THE VANDALS

Whether or not it was a true crisis, the energy crunch of 1973–74 created a sense of vulnerability in the U.S. that was to have profound implications. Almost immediately, it sparked a major rethinking about the role of oil in American national security. That rethinking would bring about a huge policy shift in which influencing and even controlling developments in the Middle East—where over 65 percent of the world's oil is located—eventually became the central focus of U.S. foreign policy. As Washington extended its control across the Middle East, both politically and militarily, the U.S. increasingly took on the role earlier performed by Britain, the dominant power in the region from the end of World War I until the late 1960s. And as the U.S. assumed that dominant role, it encountered a backlash from the people in the region, just as Britain had. Ultimately, the strong U.S. presence, including the stationing of more than five thousand U.S. troops in Saudi Arabia, provoked a reaction that led to the emergence of a militant force led by Osama bin Laden, thereby, ironically, fulfilling fears that actions in the Middle East—in fact, U.S. actions—could end up threatening the national security of the United States.

———

In a sun-filled restaurant in an upscale neighbourhood on the outskirts of Washington, D.C., James Akins clearly knows his way around a Thai menu and handles chopsticks without a slip. He displays none of the American-centred attitudes, either in his eating habits or in his view of the world, that seem common among high-ranking American officials—a group he certainly once belonged to. After diplomatic postings to Kuwait and Iraq in the late 1950s and early 1960s, Akins returned to Washington and became the State Department's leading oil expert, just as the rising nationalism of Libya and the Middle East oil states was moving oil onto Washington's front burner. Akins was then brought into the White House after the 1972 election to develop the first comprehensive U.S. energy plan. His plan ended up putting a much greater emphasis on conservation than the Nixon White House was prepared to countenance. Akins recalls top Nixon aide John Ehrlichman (of Watergate fame) reminding him: "Mr. Akins, you've got to understand that conservation is not the Republican ethic." Still, Akins's knowledge of oil and the Middle East (he had become fairly fluent in Arabic) made him invaluable to Washington, and he was appointed ambassador to Saudi Arabia just as the energy crisis was exploding.

Akins's interest in energy conservation wasn't his only flaw as far as official Washington was concerned. Even more problematic, it turned out, were his notions about how the U.S. could best advance its interests in the Middle East. In the lingo of the State department, Akins was considered an "Arabist"—that is, an expert knowledgeable in Middle Eastern culture and history. (Prominently displayed in his home near the restaurant are shelves of ancient pottery he and his wife acquired in the early 1960s in Iraq.) In recent years Arabists have all but disappeared from the State Department and the rest of the Washington bureaucracy, replaced by those with a strong focus on Israel.

Being an Arabist might imply a certain sympathy for an Arab point of view, but Akins was clearly urging policies that he regarded as being in the long-term interests of the United States.

He thought Washington was needlessly making enemies in the Arab world with its hardball approach and its refusal to accept Arab actions even when those actions were perfectly lawful and understandable. So, for instance, Akins had tried to convince Washington to drop its ban on U.S. officials talking to or having dealings with OPEC in the 1960s. "I said OPEC's got to be recognized." Similarly, as mentioned in an earlier chapter, Akins advised those in Washington that Qaddafi's demand for a forty-cent-a-barrel royalty increase in 1969 was not unreasonable, even though the demand had enraged the international oil companies. Akins actually encouraged the oil companies to accept the Libyan demand—advice the companies ignored.

Akins's advice on these sorts of matters was frequently ignored, both in Washington and among oil industry executives, with whom he often dealt. He recalled a very senior oil company official visiting his office in Washington in the late 1960s and predicting that Saudi Arabia would be producing more than 20 million barrels of oil a day by 1980. When Akins asked what the Saudis thought of such a huge daily output from their reserves, the reply was quick and dismissive: "What do the Saudis have to do with it?" It was just this sort of attitude, Akins considered, that was ultimately undermining U.S. interests in the Middle East.

American attitudes towards the Arab world hardened further after the oil shocks of 1973–74, and Akins was soon to discover just how high up the chain of command those antagonistic anti-Arab feelings could be found. By the mid-1970s the popular press, as well as academic political journals, were full of articles advocating a more confrontational approach to the OPEC challenge. Prominent conservative strategist Robert Tucker made the case for a militaristic response in an article provocatively called "Oil: The Issue of American Intervention" in the January 1975 issue of the influential journal *Commentary*. Upping the ante further, *Harper's* magazine ran a piece in its March issue bluntly titled

"Seizing Arab Oil," written by Miles Ignotus, identified in the magazine as a pseudonym for a Washington-based professor and defense consultant with ties to high-level policy-makers. The piece was full of vitriol against the Arab oil states, accusing them of extortion and blackmail. In the event of another oil embargo, Ignotus laid out a scenario in which Washington would send in two U.S. Marine Corps divisions totalling forty thousand men, supported by aircraft carriers, destroyers and ten nuclear submarines, to seize crucial oil fields in places like Kuwait and Dubai, and then bring in Americans to run them. Defence analyst Edward Luttwak has claimed authorship of the article.

But Akins believes the scenario laid out in the article came from Henry Kissinger, who was the secretary of state. Akins notes that the article is strikingly similar, in its premises and recommendations, to a half-dozen other articles that appeared at the same time. "It was clear that a senior U.S. government official had given a 'deep background' briefing." This meant that those present were forbidden to attribute anything to the briefer, but could use the information and analysis and their own. Akins was horrified by the suggestion that the U.S. should seize Arab oil fields. When he was asked on American television what he thought of the *Harper's* article, he suggested that the author must be either "a madman, a criminal or an agent of the Soviet Union." Akins now believes his remarks must have offended someone very high up in the White House. "I've been told by people at the briefing that Kissinger was the briefer." Shortly after this, Akins was fired from his Saudi posting.

Almost thirty years later, the tall, slim, casually dressed Akins fits easily into the surroundings in the low-key Thai restaurant, where he appears to be a regular client. He gives little indication of being someone who's dealt one-on-one with Saudi kings, U.S. presidents and heads of the world's most powerful corporations. He has no contact with the current U.S. administration, which never seeks his counsel, despite the depth and breadth of his grasp of two crucial subjects that remain at the centre of U.S. foreign policy: oil and the Middle East. Akins is clearly not impressed with the Bush administration's policy on Iraq, but he sees a continuity with the past. "The plan to take over Iraq is a revival of an old plan that first appeared in 1975," he says as he spears a slithery piece of eggplant in one of the exotic Thai dishes. "It was the Kissinger plan. I thought the whole idea had been killed."

Not killed, it turned out. On the contrary, enhanced. Instead of 40,000 Marines seizing the oil fields of a few unarmed, unpopulated, politically marginal sheikdoms, some 140,000 U.S. troops were sent in to seize control of a country of 25 million people at the very heart of the Arab world.

The crucial importance of oil had been amply driven home to Western nations as early as the First and Second World Wars. The British had come to appreciate the strategic importance of oil in powering the British navy in World War I. For the U.S., the pivotal lesson had been learned in World War II. Daniel Yergin notes that, by 1942, Washington was developing a whole new outlook, centred on the importance of oil. "Oil was recognized as the critical strategic commodity for the war . . . If there was a single resource that was shaping the military strategy of the Axis powers, it was oil. If there was a single resource that could defeat them, that too was oil." All this created a new sense of urgency about securing access to the vital resource. And that, of course, led to a focus on a particular part of the world. As Herbert Feis, economic adviser to the State Department, put it: "In all the surveys of the situation . . . the pencil came to an awed pause at one point and place—the Middle East." This explains why Franklin Roosevelt, on his way home from the 1945 post-war planning conference in Yalta, stopped in the Middle East for his historic meeting with Saudi king Ibn Saud.

So the strategic importance of assured access to oil—and therefore the Middle East—has been very much on Washington's agenda since World War II; the energy crisis of 1973–74 heightened that focus and put the issue front and centre before the American people. The notion that America could be deprived of its oil lifeline was now a hot political issue, reinforcing Washington's determination to ensure future access to oil supplies.

While the desirability of gaining control over the Persian Gulf had long been appreciated—and was reinforced by the energy crisis—the sheer logistics of the task deterred Washington in the 1970s. With more than 250,000 square miles of desert, 6,000 miles of pipelines and 400 pumping stations, the practical problems for troops trying to secure the area were understood to be immense. "The military demands for an operation of this size would be far beyond the capabilities of any country—or group of countries—on earth," noted Senator Mark O. Hatfield. In addition to the daunting physical geography involved, there was the reality of Soviet military power and the likelihood that the Soviets would try to prevent the U.S. from taking control of a crucial oil region, and one that was right in their backyard.

Within a few months of the article in *Harper's*, the U.S. Congressional Research Service produced a study assessing the prospects for a U.S. occupation of the Gulf, and highlighting the difficulties: seizing oil installations intact, securing them (possibly for years), operating them without the owners' assistance, and guaranteeing safe passage overseas of supplies and petroleum products. The study concluded that a successful outcome of an occupation would be possible only if there were minimal damage to oil installations and no Soviet armed intervention.

Since neither essential could be assumed, military operations to rescue the United States (much less its key allies) from an airtight OPEC embargo would combine high costs with high risks . . . This country would so deplete its strategic reserves that little would be left for contingencies elsewhere. Prospects would be poor, with plights of far-reaching political, economic, social, psychological, and perhaps military consequences the penalty for failure.

It's interesting to note that, while the difficult logistics of such an operation were considered grounds for caution, there seemed to be little questioning of the moral justification for an occupation. The prevailing attitude, even at the highest levels, was that the Arab countries had committed a terrible transgression by denying the West access to their oil and that the West was justified in taking strong measures to restore that access. Asked if Washington would consider military action in the case of another round of price increases, Kissinger told *Business Week* that it was one thing to face price increases (which could be negotiated), but "it's another where there is some actual strangulation of the industrialized world." Kissinger's comment provoked outrage in the Arab world, but the U.S. didn't back off or even try to soften the tone. Rather, President Gerald Ford firmly supported Kissinger's position, telling *Time* magazine a few weeks later, "I wanted it made as clear as I possibly could that this country in case of economic strangulation—and the key word is strangulation—we had to be prepared, without specifying what we might do, to take the necessary action for self-preservation."

Secretary of Defense James R. Schlesinger continued the sabre-rattling when he told *U.S. News and World Report* in May 1975 that "we might not remain entirely passive to the imposition of [another oil] embargo. I'm not going to indicate any prospective reaction, other than to point out that there are economic, political, and conceivably military measures in response." Of course, in all these remarks, the U.S. was depicted as a wounded party, an innocent victim simply trying to protect itself against the aggression of others. James H. Noyes, deputy assistant defense secretary, took this image of American innocence and vulnerability to new heights when he argued that U.S. military action in the Gulf would be justified because OPEC's massive revenue grab risked creating "a somewhat impoverished America surrounded by a world turned into a slum."

Despite this apparently burning sense of justification, Washington held back. It was still recovering from the Vietnam debacle, in which the long and bloody U.S. military intervention had ended in a thinly disguised defeat (although it was never officially acknowledged as such). Rather than risk U.S. troops being involved in another major foreign fiasco, Washington developed a strategy of relying on local powers—Iran and Saudi Arabia, but particularly Iran—to advance Western interests in the region. By selling billions of dollars' worth of weaponry to these dictatorships, Washington was hoping to fill the power vacuum that followed the withdrawal of the British. "What we decided," Undersecretary of State Joseph Sisco explained later, "is that we would try to stimulate and be helpful to the two key countries—namely, Iran and Saudi Arabia—[so] that . . . they could become the major elements of stability as the British were getting out."

But the role of Iran and Saudi Arabia as surrogates for U.S. power tied these countries more tightly to Washington, fuelling further resistance within their domestic populations. In Iran, that resistance eventually led to the overthrow of the shah through a massive popular uprising. After being overwhelmingly rejected by his own people, the shah was permitted to go to the United States, demonstrating once again Washington's friendship with the hated dictator. Shortly afterwards, in November 1979, a group of militants stormed the U.S. embassy in Tehran and seized fifty-two hostages. The hostage-taking drama, which dragged on for fourteen months, was pivotal in hardening the attitudes of ordinary Americans towards the Islamic world. As American television was filled day after day with images of Iranian militants chanting anti-

U.S. slogans—including a whole new nightly news program called *America Held Hostage*—the American public was left apparently baffled by the strange, menacing young men on their TV screens who seemed out of nowhere to have decided to pick on the United States. "Americans," observed *New York Times* correspondent Stephen Kinzer, "found this crime not only barbaric but inexplicable. That was because almost none of them had any idea of the responsibility the United States bore for imposing the royalist regime that Iranians came to hate so passionately."

With the fundamentalist, anti-U.S. cleric Ayatollah Ruhollah Khomeini now running Iran and militants openly challenging U.S. power, it was abundantly clear that Washington could no longer rely on that country to promote its interests in the Middle East. "[I]t was evident to senior policy-makers that the surrogate approach was no longer viable and that, as a consequence, the United States would have to assume *direct* responsibility for stability in the Gulf," writes political scientist Michael Klare, author of *Resource Wars*. Accordingly, Washington moved a step closer to establishing U.S. military control over the area. In his state of the union address in January 1980, President Jimmy Carter declared that "an attempt by any outside force to gain control of the Persian Gulf region will be regarded as an assault on the vital interests of America, and such an assault will be repelled by any means necessary, including military force." Although put in the language of defense—protecting the region from an outside attacker—the Carter doctrine clearly laid out for the first time that not only did the U.S. consider the area vital to its interests, but it would use direct military force to ensure access.

Thus began in earnest a reorientation in U.S. military strategy that brought the Middle East clearly into the crosshairs. Key to this was the establishment of the Rapid Deployment Force (RDF), a flexible, quick-response force that was set up to carry out emergency military tasks in situations involving less than all-out nuclear war. While the RDF had actually been around in various forms since the early 1960s, it only received substantial budgetary support after 1980, with the hostage-taking in Iran in November 1979 and the Soviet invasion of Afghanistan six weeks later. By 1981, Washington's long-standing preoccupation with NATO and the Soviet threat to Europe had undergone a significant change, notes U.S. military analyst Jeffrey Record, as the focus—and the funding—shifted towards creating the capacity for quick-strike military interventions elsewhere, particularly the Middle East. In 1983, the RDF was expanded and renamed the U.S. Central Command (CENTCOM), with its primary focus being the defense of the Gulf region from a possible Soviet invasion.

Even with ample new financing for the force, the strategic difficulties involved in gaining control over the Gulf remained immense. For one thing, there was still the problem of the Soviet Union, which, it was assumed, would intervene to prevent an American takeover of the region. For another, there was the fact that most of the troops in this well-funded new force were located on bases in the U.S., thousands of miles from the area. Washington moved to overcome this problem, establishing a permanent naval force in the Gulf, as well as upgrading its existing military bases there and acquiring new ones. But establishing more of a land-based military presence in the region was challenging, largely because governments feared that this would not be well regarded by their populations. So, for instance, the Gulf state of Oman was willing, for $100 million, to grant Washington access to its airfields and port facilities located near the crucial Strait of Hormuz. (The strait has been described by the U.S. Energy Department as "the world's most important oil checkpoint," through which some 15 million barrels of oil pass each day.) But there were severe restrictions placed on the U.S. forces deployed in Oman, presumably to prevent them from being too visible to the local population. In one 1981 military exercise, a

Marine Corps landing force "was allowed to penetrate only four miles on the Arabian sea coast and remain on Omani territory for only thirty hours."

As it extended its military reach over the Gulf, Washington was clearly brushing up against the sensitivities of local populations. But it faced little meaningful opposition from leaders in the region—with one exception.

———

By the early 1990s, Saddam Hussein had become the globe's foremost villain in the U.S. world view. But Washington's strained relations with Iraq actually predated Saddam. In 1958, Brigadier General Abdul Qarim Qasim had taken power after he led a military coup, overthrowing the British-installed monarchy that had ruled Iraq since the end of World War I. Qasim's coup was generally well received by the Iraqi population, who welcomed the end of British control. Qasim went on to be a fairly popular leader, particularly among the lower classes, according to Stephen Pelletiere, a professor at the U.S. Army War College and the author of *Iraq and the International Oil System*. Qasim introduced a number of social reforms, including land redistribution, social insurance and unemployment programs, and subsidies to keep bread prices down. He also adopted a tough approach to the foreign oil companies operating under the umbrella of the Iraq Petroleum Company (IPC). Qasim pushed hard for increased production and pressed IPC to give Iraq a 20 percent stake in the company.

Much to the dismay of the international oil industry and Washington, Qasim's regime appeared to be heading in the same nationalistic direction that Mossadegh had pursued in Iran earlier in the decade. Relations became even more tense in December 1959 when Qasim, frustrated by IPC's failure to cooperate with his requests, increased shipping duties on the company. In retaliation, IPC shut down production entirely at the Rumaila oil field and cut back production at other Iraqi fields. With tension mounting, Qasim passed the controversial law, mentioned in Chapter 7, taking back most of IPC's concession—a move which, Pelletiere notes, enjoyed popular support but created lasting resentment among the international oil companies that controlled IPC.

In 1963, Qasim was overthrown in a violent coup led by the Baathist party—a hierarchical, anti-Communist organization backed by street toughs, including a young Saddam Hussein, who was injured in the violence. The next few years were marked by instability until a second group of Baathists, including an increasingly powerful Saddam, managed to firmly take power in 1968. There are credible reports of CIA involvement in the Baathist power grabs of both 1963 and 1968, according to Pelletiere, who later worked as a CIA operative himself in the region. But Pelletiere argues that it isn't possible to conclusively substantiate these reports, although other analysts disagree. What is clear is that, after the Baathists consolidated their power in 1968, with Saddam effectively taking control of the party, they moved Iraq in a direction regarded as hostile by the West.

Saddam followed the nationalistic approach initiated by Qasim, moving even more aggressively in this direction. The state-owned Iraqi National Oil Company, which Qasim had established but kept on the sidelines, was activated under Saddam, and became involved in exploration, development and future planning. While Saddam's brutal and repressive police-state methods were hated by Iraqis, his nationalistic approach to oil had widespread support, as did his investment of oil revenues in the construction of an impressive public infrastructure. In 1972, Saddam began nationalizing Iraq's oil fields, a process completed in 1975. "In the eyes of the companies, this was an unpardonable crime," observes Pelletiere. Even worse, in the eyes

KING OF THE VANDALS   299

of Washington, were Saddam's increasingly close ties with Moscow. In 1974, Saddam signed a Friendship Treaty with the Soviets, to whom the Iraqi dictator turned, not only for help in establishing the apparatus of a totalitarian state, but also for assistance in developing his oil infrastructure.

Washington's hostile relations with Saddam then took a curious turn, becoming considerably warmer after the outbreak of the Iran–Iraq War in 1980. Saddam was still his same bloody, barbarous self, crushing internal resistance and maintaining a hammerlock on the Iraqi state. But the Americans now found him useful in his willingness to fight the Islamic fundamentalist regime that had taken over in Iran and had allowed young militants to hold U.S. citizens hostage in Tehran for so many months. Hence Washington, while officially neutral, actually aided Saddam during the Iran–Iraq War. (In fact, Washngton also secretly sold arms to Iran—a fact that was revealed during the Iran–Contra scandal which engulfed the Reagan administration in 1986.) U.S. assistance for Iraq, which was highly classified, involved more than sixty officers of the U.S. Defense Intelligence Agency, who provided Iraq with intelligence on bomb damage assessments inside Iran as well as satellite photos of Iranian troop deployments. Washington also facilitated Iraq's loan requests and lobbied allies to cut off arms supplies to Iran. And in 1984 the U.S. withdrew its objections to Saddam taking possession of fighter jets Iraq had purchased from France, allowing the missile-equipped planes to be delivered. The U.S. support for Saddam in those years is, of course, particularly intriguing given Washington's later fierce determination to oust the Iraqi dictator. Moreover, Washington's support during the 1980s was offered despite the knowledge that Saddam was using chemical weapons in the conflict.

Most intriguing is the role played at the time by Donald Rumsfeld, who later went on to oversee the U.S. invasion of Iraq in 2003 as defense secretary. Twenty years earlier, in 1983,

Rumsfeld, then head of a pharmaceutical company, went to Baghdad as a special envoy of the Reagan administration to promote closer U.S. relations with Saddam. (Photographs capturing the apparently pleasant encounter between Rumsfeld and Saddam surfaced in the media in the run-up to the 2003 U.S. invasion, and have remained popular on Internet sites.) Rumsfeld's second trip to Baghdad a few months later, in March 1984, is even more rife with ironies. Saddam's use of chemical weapons in the war against Iran, and also against Kurdish rebels, had been widely reported and criticized, prompting the Reagan administration to issue a public condemnation of Iraq on March 5. A few days later, Secretary of State George Shultz met with an Iraqi diplomat to reassure him that Iraq shouldn't regard Washington's harsh words as a rebuff. The diplomat left the meeting "unpersuaded," according to documents. So Shultz, obviously regarding the friendship with Saddam as very important, asked Rumsfeld to return to Baghdad, essentially to assure Saddam that Washington's official condemnation was just for show. In a March 24 briefing document, Rumsfeld was told to clarify Washington's position "that our CW [chemical weapons] condemnation was made strictly out of our strong opposition to the use of lethal and incapacitating CW, wherever it occurs." But Washington wanted it "emphasized that our interests in 1) preventing an Iranian victory and 2) continuing to improve bilateral relations with Iraq, at a pace of Iraq's choosing, remain undiminished." Rumsfeld was told: "This message bears reinforcing during your discussions."

U.S. support for Saddam evaporated soon after Iraq emerged victorious from its war against Iran. With that 1988 victory, Saddam's power was considerably enhanced in the region—a matter of some concern to Washington. Indeed, with the war over, Washington returned to its previous hostility towards Saddam. After all, he was a Soviet ally, and he was also the most fiercely anti-Israel of the Arab leaders; he maintained close ties with the Palestine Liberation Organization and actively championed its demand for a Palestinian state, when other Arab states had backed off that cause. Also worrisome to Washington, no doubt, was the likelihood that Saddam, now strengthened by victory and well equipped with sophisticated French and Soviet weaponry, would try to oppose U.S. attempts to extend its military influence over the region. Of all the states in the Gulf area, Iraq had been the most resistant to Washington's efforts to establish a strong military presence. According to Pelletiere, Saddam "wanted to preclude the United States from coming into the Gulf, and in this he had the backing of the [Gulf] sheiks, who did not want their part of the world to become an arena of superpower rivalry." Saddam declared an Arab Charter asserting that the Gulf states themselves would protect the Gulf waters, rather than allow that role to fall to Washington. (Saudi Arabia declined to back Saddam's charter, but four Gulf states supported it.)

So a confrontation between the U.S. and a more assertive, confident Iraq was perhaps inevitable, with hegemony over the vital oil-rich nations of the Gulf region at the centre of their rivalry. By late 1989, General Colin Powell, then chairman of the Joint Chiefs of Staff, was ordering General Norman Schwarzkopf, commander-in-chief of CENTCOM, to draw up a blueprint for a major military confrontation with Iraq. After five months, such a plan was outlined in a document, *Operational Plan* [OpPlan] *1002–90*, that, as reported in the media, called for the deployment of several armoured divisions, backed up by air and naval forces. CENTCOM even conducted a dry run of an imagined Iraq–U.S. conflict in mid-July 1990. Two weeks later, as if on cue, Saddam invaded Kuwait.

Saddam's decision to invade Kuwait remains puzzling. Iraq had always maintained that Kuwait should be part of Iraq, but the two countries had had relatively good relations during the Iran–Iraq War, when Iraq took on the role of protector, shielding Kuwait—a fellow Arab nation—from possible aggression by non-Arab Iran. Kuwait showed its gratitude (along with Saudi Arabia) by providing financial help for Iraq's war effort. After the war, however, relations soured. Iraq tried to get Kuwait to cede two small islands that were crucially located right at the mouth of the narrow waterway that was Iraq's only access to the Persian Gulf. The islands were not crucial to Kuwait, which has ample coastline on the Gulf. Nevertheless, Kuwait refused to hand them over, offering instead to lease the islands to Iraq for billions of dollars. And Kuwait also demanded repayment of its huge war loans, which Iraq had assumed were grants. To make matters even more contentious, Kuwait was cheating on its OPEC quota, producing too much oil and consequently bringing down the price for everyone—including Iraq, which was badly in need of money.

An Iraqi takeover of Kuwait would certainly have given Iraq the control it wanted over access to the Persian Gulf, and potentially solved its financial problems too. What is puzzling, however, is Iraq's apparent belief that it could invade oil-rich Kuwait without provoking the U.S., when Washington clearly considered the whole region part of its domain. Did Saddam really think he could invade Kuwait—right next door to Saudi Arabia—without bringing the U.S. Army down on him? One possible explanation is that Saddam believed the U.S. had indicated it did not object to the invasion, or at least did not intend to interfere with it.

Saddam might possibly have drawn this conclusion from his meeting with U.S. ambassador April Glaspie, whom he took the unusual step of summoning to his presidential palace on July 25, 1990, a week before he launched the invasion. According to an official Iraqi transcript of the encounter, which was later reported in the Western media, Glaspie questioned why Saddam had so many troops massed at the Kuwait border. Saddam responded that, if he couldn't get Kuwait to agree to his demands, he intended to move against Kuwait. Then he asked Glaspie for Washington's opinion on the matter. "We have no opinion on your Arab-Arab conflicts, such as your dispute with Kuwait," Glaspie is quoted as saying in the Iraqi transcript. "Secretary [of State James] Baker has directed me to emphasize the instruction, first given to Iraq in the 1960s, that the Kuwait issue is not associated with America."

Is the Iraqi transcript accurate? British journalists got a copy of it about a month after the meeting—after the invasion had been launched—and confronted Ambassador Glaspie as she left the U.S. embassy in Baghdad. Although she mostly remained silent, pushing past the journalists towards her limousine, she did respond to a question about whether she had encouraged the Iraqi aggression against Kuwait. "Obviously, I didn't think, and nobody else did, that the Iraqis were going to take *all* of Kuwait," she said. Months later, in testimony before the U.S. Senate Foreign Relations Committee, Glaspie described the transcript as a "fabrication" that distorted her position, although she conceded it contained "a great deal" that was accurate. She was later demoted from ambassador to the rank of consul general, serving in Cape Town, South Africa.

It's interesting that Washington has never taken steps to disprove the accuracy of the Iraqi transcript by, for instance, releasing State Department instructions to Glaspie prior to the interview. In 1992, presidential challenger Ross Perot asked President George H.W. Bush during a televised debate why he wouldn't at least allow the written instructions to Ambassador Glaspie to be released to the Senate Foreign Relations Committee or the Senate Intelligence Committee. Bush brushed aside the suggestion: "That gets to the national honour! That is absolutely absurd." In any event, Washington has never denied that Glaspie's meeting with Saddam took place, nor that the U.S. was aware of the buildup of Iraqi troops on the Kuwaiti border. Presumably this troop deployment would have prompted some kind of negative reaction from the U.S., whether from Glaspie or from officials in Washington. Washington is not generally shy about letting countries know what it would like them to do—or not do. And yet, curiously, there was no public statement from the U.S. warning Iraq not to enter Kuwait.

If the Iraqi transcript *is* accurate, it raises the question of whether the U.S. deliberately avoided expressing any opposition to the invasion—in fact, signalling active indifference—so that Iraq would proceed into Kuwait, thereby handing Washington the perfect justification for U.S. military intervention. With the collapse of the Soviet Union, it was now possible for Washington to intervene without fear of massive retaliation. As it turned out, there was even international support for such a mission against the Iraqi aggressors. So, if the transcript is accurate, it suggests that Washington may have sought to encourage the Iraqi invasion in order to pave the way for full U.S. military intervention. Certainly, in the war that followed, things unfolded pretty much as Schwarzkopf had set out in OpPlan 1002–90. Iraq's rising military might, which had made it the dominant military power in the area, was utterly crushed by the superior air and ground forces of the U.S. and its allies. Thus, the only rival of U.S. power in the region was destroyed as a military force. In the aftermath of the war, thousands of U.S. troops remained in the area, stationed in Saudi Arabia and Kuwait.

The U.S. had finally achieved a significant—indeed dominant—military presence in the Gulf. And there'd been barely a murmur of international protest.

————

If there had been any ambiguity about U.S. military intentions in the Gulf before the 1991 war, there was none after it. "The U.S. will continue to use a variety of means to promote regional security and stability [in the Gulf], working with our friends and allies . . . and will remain prepared to defend vital U.S. interests in the region—unilaterally if necessary," stated Assistant Secretary of

Defense Joseph Nye in 1995. Iraq itself was kept largely incapac-
itated, with the UN (at U.S. insistence) imposing far-reaching
sanctions that kept out supplies and equipment of all sorts, even
those with no conceivable military uses. Washington established
no-fly zones over the north and south of the country, and
enforced them with U.S. (and British) aircraft operating from
bases in Saudi Arabia and Turkey. Washington also provided
more than $42 billion worth of weaponry, ammunition and
extensive military training to Gulf states, including Saudi Arabia,
Kuwait, Bahrain, Oman and the United Arab Emirates, with the
idea of equipping them, not to operate on their own, but so "their
forces could make a distinct contribution to a joint effort with the
United States," according to a Strategic Assessment produced by
the Pentagon's Institute for National Security Studies in 1998.
The intention of the U.S. to control the area could hardly have
been clearer. "All these endeavours," notes Michael Klare, "are
part of a consistent, integrated policy aimed at bolstering
American military dominance in the Persian Gulf area. Indeed,
securing such dominance in the Gulf has been one of the most
important and persistent goals of American strategy since 1980."

What is also amply clear, despite insistent attempts by some
commentators to deny it, is that this U.S. push for military domination of the Gulf has always had a lot to do with controlling the
region's oil reserves, which, as Cheney has pointed out, amount
to roughly two-thirds of the world's known supply of this most
precious of all resources. This focus on oil was acknowledged in
the report *The Geopolitics of Energy into the 21st Century*, which was
prepared in November 2000 with the help of a bipartisan congressional team, administration officials and oil industry
executives. The carefully worded report made clear that
Washington's stated purpose of liberating Kuwait was only part
of its motivation: "Although the primary objective of the United
States and the allied forces came to be the liberation of Kuwait

and the restoration of its sovereignty, the issue of energy security was also a powerful motivating force."

With the arrival of the George W. Bush administration in 2001, the U.S. interest in gaining military control over the Gulf took a new, more aggressive turn. The neo-conservatives who formed the core of the new administration had been pressing throughout the 1990s for the overthrow of Saddam. With the Soviet threat gone and Saddam's army effectively destroyed, such an intervention was now tantalizingly possible in a way that wasn't the case when these neo-cons were last in positions of power, during the Reagan and Bush Sr. administrations. There was no power that could stop Washington now. Although Bush had never mentioned such an intervention in his election campaign, it immediately became a top-priority issue for his administration, as the account of former Treasury secretary Paul O'Neill has made clear. Seizing direct control over some of the richest oil fields of the Middle East—a plan first considered more than a quarter of a century earlier—was now actively in play.

———

Those who scoff at the notion that the invasion of Iraq was motivated by an interest in oil often suggest that there would have been other ways to get the oil. Michael Ignatieff, a Canadian academic who teaches at Harvard University, has made this sort of argument: "If all America cared about was oil, it would have cozied up to Saddam," thereby presumably assuring America's access to Iraq's oil. But *access* to oil is different from *control* over oil. Without control, access can be cut off at any time. With control comes not only inevitably bigger financial rewards but also mastery of the world's most crucial resource. Control has therefore always been the ultimate goal. This is why the insightful oil analyst and historian John M. Blair titled his 1976 book—hailed in a front-page review in *The New York Times Book Review* as the "definitive book on oil"—*The Control of Oil*.

The desire for control, as opposed to simply access, was clearly demonstrated with the U.S. intervention in Iran in 1953. In nationalizing its oil industry, Iran was in no way attempting to deny the West *access* to its oil. Quite the contrary. Iran was very much hoping to sell its oil to the West and to the rest of the world, and it was willing to take whatever price the market would bear. But Washington and London (and the international oil companies) weren't satisfied with merely being able to buy unlimited amounts of Iranian oil; they wanted Iranian oil to be, once again, under the control of the international oil companies.

As the oil producing nations became more militant and organized in the 1970s, it was no longer possible for the West to control oil as it had before. This loss of control was a tremendous blow to Washington, particularly when the Arab nations acted together to actually cut back their oil shipments to the U.S. in 1973–74. Although the Arab oil embargo was brief, it had a huge psychological impact on the U.S., prompting Washington over the next few decades to greatly build up its military presence in the region in order to prevent America's economy and choice of lifestyle from ever being put at risk like that again. Thirty years later, that embargo remains a potent memory in Washington—one of the rare moments in U.S. history when the country felt vulnerable, unable to prevent those outside its borders from inflicting serious harm on the nation.

The desire to shield itself from a future oil embargo has no doubt taken on new meaning in Washington in light of its escalating tensions vis-à-vis the Islamic world after 9/11. By 2010, an estimated 95 percent of the remaining global oil export capacity will be located in the Muslim world—a troubling fact only if one expects the West to be embroiled in endless, bitter conflict with the world's Muslims. Many in the Bush administration seem to see such conflict as inevitable and unavoidable—a prophecy that may come true if the U.S. continues to refuse to deal substantively with one of the most potent sources of

Muslim anger against the West: the deferred establishment of a Palestinian state. But then, if the goal is ultimately to deny that statehood, an ongoing bitter conflict between the West and Islam may be seen as useful, in that it reinforces an us-against-them mentality.

Certainly, control over oil is likely to become more important as easily accessed oil reserves around the world are depleted and competition for the remaining reserves becomes more intense. In such a scramble, controlling oil will be a strategic lever in the global power game, a necessary tool for a superpower hoping to maintain its global dominance. "The U.S. views oil as a key weapon," says oil analyst Michael Tanzer. If so, Ignatieff's suggestion that Washington could have achieved all it wanted by cozying up to Saddam seems to miss the point. Why would the U.S. want its fate left up to the whims of Saddam, a treacherous, anti-Western autocrat who shared none of Washington's political goals or ideology? Why would the most powerful country on earth want to be obliged to rely on some two-bit regional dictator, or whoever might replace him? If oil is a key weapon, then it's a weapon that Washington wants in *its* arsenal, not someone else's.

———

When Genseric, king of the Vandals, invaded northern Africa in A.D. 428, he probably didn't declare that his intention was to plunder and pillage. He *did* plunder and pillage, though, which is why, some sixteen centuries later, the name of his people has ended up as an enduring word in our vocabulary, synonymous with thuggery and hooliganism.

Invading armies are often coy when it comes to admitting their true motives. Certainly the desire to seize territory or resources is rarely among the motives that modern invading armies tend to highlight. One can understand the preference for looking like a liberator rather than a pillaging bully.

What is harder to understand is how members of the media step forward to make the invaders' case for them. As violence in Iraq escalated in the fall of 2003, Thomas Friedman enthused in *The New York Times* that "this is the most radical-liberal revolutionary war the U.S. has ever launched—a war of choice to install some democracy in the heart of the Arab-Muslim world." The whole twenty-five-year history of planning to take military control of the Gulf in order to secure the oil supply had been airbrushed out. Even the claims that the war had been motivated by fears that Saddam possessed weapons of mass destruction and had ties with Al Qaeda had been conveniently set aside. Instead, the latest Bush administration claim—that the war was actually about installing democracy in the Middle East—was now presented without qualification or hesitation as the true motive, and simultaneously heralded as a giant step in the direction of democratic humanism. Whoa, baby.

Bob Woodward has helped flesh out this portrait of George W. Bush as a determined liberator of the oppressed. Promoting his blockbuster best-seller *Plan of Attack* in April 2004, Woodward expressed the view that Bush's motivation for the Iraq invasion was "his duty to free people, to liberate them." What can one say about Woodward? It's hard not to like the guy. Popping up on virtually every major TV interview program throughout the month of April, he came across as a genuinely well-meaning, caring sort of person, someone you'd like to have as a friend. One senses a certain honesty about the man. One also, however, detects a certain gullibility. He seems to believe what Bush tells him, to trust the president. (He also appears to have become almost chummy with the president. After the publication of his 2001 book *Bush at War*, Woodward told an interviewer that he'd spent quite a bit of time with Bush, and that the president had even given him the nickname Woody.) Woodward routinely goes beyond merely reporting Bush's remarks; he weighs in with his opinion about the sincerity of Bush's motives, offering his view that the president decided to invade Iraq because of "his duty to free people."

Needless to say, this explains why Woodward is given so much access to Bush and the rest of the White House team; the White House couldn't ask for a better vehicle than Woodward to get across its message. After all, he's an icon of American journalism, an investigative reporter who (along with Carl Bernstein) unravelled and exposed the entire Watergate scandal and in the process brought down Richard Nixon. Not only are Woodward's past journalistic credentials impeccable, but he seems doubly believable because he doesn't appear to have a point of view. Indeed, one suspects he thinks that the invasion of Iraq was a mistake; he certainly provides a lot of evidence that can be used to bolster the case that it was. But the bottom line is that Woodward presents the president as being sincere in his motives, and thereby lends his own enormous journalistic credibility to the portrait of George W. Bush as a man of integrity and principle.

Of course, it's possible that Bush *was* primarily motivated by a desire to liberate people, that he wasn't much concerned about Washington's long-time strategic interest in gaining military control over the Gulf. Bush actually tried to distance himself from past U.S. policies in the region, suggesting he intended to focus more on encouraging democracy and freedom. He even appeared to be willing to acknowledge that the West had supported repressive regimes in the past, but he said he intended to change that. Addressing the National Endowment for Democracy in November 2003, Bush declared: "Sixty years of Western nations excusing and accommodating the lack of freedom in the Middle East did nothing to make us safe."

The following month, however, Donald Rumsfeld visited Baku, the capital of Azerbaijan, an oil-rich Muslim nation carved out of the old Soviet empire, just north of Iran.

Azerbaijan had just sworn in a new president, Ilham Aliyev, who had inherited the presidency from his father, a corrupt and tyrannical dictator. Like father, like son, it seems. The younger Aliyev took power after an election denounced by international observers as fraudulent. More than a thousand protestors had been arrested, including journalists, opposition leaders and election officials; a report by Human Rights Watch documented cases of beatings and torture. But, for all its democratic deficits, Azerbaijan was considered useful to the U.S., both as an ally in the "war on terrorism" and also for its significant reserves of oil. Over the past decade Aliyev and his father had granted billions of dollars in oil contracts to ExxonMobil, ChevronTexaco and BP-Amoco, and had supported a crucial $3-billion pipeline to transport oil from the Caspian Sea to a port in Turkey.

It's hard to resist noting a few ironies. Here was Donald Rumsfeld—who had cozied up to brutal Saddam on behalf of the U.S. in the early 1980s and twenty years later led the U.S. invasion that overthrew him—now apparently cozying up to the brutal dictator of yet another oil-rich, strategically located state. Asked by reporters about allegations of electoral fraud by Azerbaijan's new leader, Rumsfeld curtly replied: "The United States has a relationship with this country. We value it."

If there'd been a change of heart in the White House about "excusing and accommodating" dictators in the Middle East, apparently the new policy didn't extend a few hundred kilometres northeast to Azerbaijan.

Still, pundits like Thomas Friedman insist that U.S. actions in the region are motivated by something more than the run-of-the-mill imperialism and plundering that motivate other invaders. Perhaps, sixteen centuries from now, the word "bush" will endure in the vocabulary, synonymous with bringing liberty to a people. But somehow I doubt it.

CHAPTER 10

VROOOOOOM!

Other than the Holocaust, probably no subject has so dominated western thought in recent decades as the events relating to September 11, 2001. Certainly no other subject has taken up so much newspaper space or air time, as the media have relentlessly explored every aspect of the horror and the tragedy, as well as repeatedly analyzing every possible consequence and ramification of the attacks and the "war on terror" unleashed in their name. It has become a commonplace to assume that the world has fundamentally changed as a result, with some even concluding that the map has been redrawn into Biblical-style realms of good and evil. Yet despite the frenzy of commentary that has surrounded 9/11, one thing was clear right from the start: there would be no meaningful discussion of what *caused* the attacks. Any probing of the so-called "root causes" would be strictly off-limits.

On one level, this seemed reasonable. An attempt to understand what motivated the perpetrators of such atrocities might give credence or even justification to their brutality. If there were "grievances" harboured by the terrorists, it seemed somehow inappropriate to discuss them now that the terrorists had created such carnage. Any such discussion risked "rewarding" them, by paying attention to the very issues they had been trying to put on the agenda. It also meant a willingness to open up discussion—and inevitably criticism—of past U.S.

government actions at a time when America was feeling very much the victim.

But while the impulse to block this debate may have been understandable, it also created a kind of wilful blindness. It meant there would be outrage and moral certainty about how bad the terrorists were, but no real understanding of what they thought and why they'd acted. As a result, of course, there would be no possibility of making the kind of changes that might prevent similar attacks. There would just be a digging in of heels, a hardening of positions, a ramping up of hostilities. There would be no attempt to understand this lethal enemy, who would simply be labelled "terrorist." No further explanations would be necessary, although there would be plenty of room for commentary about the evil he embodied.

It should be noted that this sort of wilful blindness isn't really a reflection of the level of evil involved. It would be hard to imagine anything more evil than Nazi Germany, and yet the "root causes" of the rise of Nazi Germany and the Second World War are routinely discussed and analyzed in history books. I remember having to answer questions about them in high school essays and exams. One of the root causes, I recall being taught, was the huge reparations bill that the Allies had forced Germany to pay after World War I—a source of deprivation, bitterness and national humiliation that had made the German people more susceptible to Hitler's vengeful cries for an assertion of German power. In explaining this background, my high school teacher was not in any way trying to let Hitler off the hook. Hitler's full criminal responsibility for Nazi atrocities was never in question. But the teacher considered it reasonable, even important, that we understand the world, and part of that involved understanding how such violence and brutality had come into play.

This urge to understand the phenomenon of Nazi Germany, and thereby to learn from it, wasn't confined to the classroom. In both intellectual and popular debate, there has been an attempt to grapple honestly and openly with the roots of what is perhaps the most horrific of all atrocities. One consequence of this attempt to learn from the past was the decision not to impose reparations on Germany and Japan after World War II; in fact, to do just the opposite—to plow billions of dollars into rebuilding these countries in order to try to entice them into joining the Western fold. Whatever the merits of this approach (it did seem to work in these cases), it is significant that it came out of an analysis of the "root causes" of the Second World War and an intense desire to avoid a similar set of circumstances from ever occurring again.

Oddly, there's been no such open-mindedness and analytical clarity—nor even, until recently, much curiosity—when it comes to the public discussion of 9/11. To the extent that the events of 9/11 are finally being probed, the questions have been almost entirely related to the failure of U.S. intelligence, or the failure of political leadership in responding to that intelligence. This important subject was the focus of the bipartisan commission investigating 9/11—a commission that the White House resisted appointing until it came under intense pressure from the families of 9/11 victims. The commission's public hearings in the spring of 2004 suddenly brought these questions—and some of the intriguing answers to them—out into the open.

Even more basic than the issue of intelligence is the question of root causes, and on this there has been a dogged refusal to open up a serious discussion. Indeed, a new political correctness has emerged—a far more effective, pervasive and comprehensive political correctness than the earlier version that (allegedly) prevented honest discussion about issues relating to women and minorities. This new political correctness has been hugely effective in discouraging meaningful debate. While there's been debate about the appropriateness of invading Iraq—with critics questioning the wisdom of the invasion and even sometimes the motives behind it—there has been virtually no serious debate allowed about the central question of why the U.S. is so hated in the Middle East and why extremist anti-American groups like Al Qaeda appear to have attracted at least some degree of popular support over there. The only answers permitted to these questions revolve around the violent nature of Islam, the irrational hatred harboured by terrorists (who are said to love death, while we love life) and the jealousy of those who lack the riches of the West. And while it's permissible to question whether invading Iraq will lead to a new round of Al Qaeda recruiting, there is virtually no serious discussion about what inspired the first round. To even suggest that there are *reasons* why Al Qaeda's message resonates with many in the Middle East is to risk censure and charges of anti-Americanism.

If there weren't such wilful blindness on this subject, there would be lots to discuss and explore. Tempting as it is simply to dismiss Osama bin Laden as a crazed madman, his grievances are fairly straightforward, even if his methods are barbaric. Bin Laden refers to "eighty years of humiliation and disgrace" suffered by the "Islamic nation." This is not particularly puzzling or hard to grasp. It is clearly a reference to the West's domination of the Islamic world, starting with the cavalier carving up of the Ottoman Empire at the end of World War I, just over eighty years ago. While bin Laden is an Islamic fundamentalist, his hostility to the West seems to be rooted in a fairly familiar phenomenon: a sense of humiliation derived from decades of foreign domination. Resentment of foreign domination—particularly by a foreign power with another language, culture and religion—is a pretty standard human response across the ages and throughout the world; it's hardly something confined to the Middle East. An unusually explosive situation has developed in that region in the last few decades as a result of a number of factors, including intense feelings among Muslims of humiliation and helplessness at the growing power of

Israel, an increasing U.S. military presence and a large amount of money available to finance a sophisticated underground terrorist network.

Oil is clearly a key part of the story. Not that bin Laden is focused on oil; his focus is clearly elsewhere—on reasserting Islamic power in the world. But the West's fervent desire for the region's oil is the beginning point of the whole narrative of western involvement. "If Iraq did not have oil, we would not give a damn about Iraq," notes Wall Street oil analyst Fadel Gheit. Similarly, if the Middle East did not have oil, it would have remained largely off the West's radar, an economic backwater attracting little more interest than, say, Africa, worthy of no more than low-priority attention. But it does have oil and, particularly after the importance of oil became clear during World War II, the West has moved to assert control over that resource. While Western oil companies had been involved in the region since the early part of the twentieth century, with Britain as the dominant military power, Washington became directly involved in March 1945 when Franklin D. Roosevelt made that detour on his way home from the Yalta Conference to meet Saudi Arabia's King Abdul-Aziz ibn Saud.

That was the beginning of the "special relationship" between the U.S. and Saudi Arabia, an agreement that tied the greatest economic and military power on earth to the biggest oil well on earth. Roosevelt offered the cash-strapped Saudi ruler $20 million plus a private DC-3 in exchange for Saudi agreement to honour oil concessions that had been awarded to two U.S. firms (SoCal and Texaco; later, Exxon and Mobil were allowed into the Aramco consortium). The exact details worked out by Roosevelt and Ibn Saud in 1945 have never been released, but the essence of their deal has long been clear and has remained in place for almost sixty years: the U.S. would be given access to Saudi oil and a say over Saudi oil policy and, in exchange, U.S. military power would maintain the Saudi royal family in power, protecting it from external and internal threats. Eventually worth trillions of dollars and backed up by the world's strongest military, it was a deal struck between a foreign power and a private interest—the Saudi royal family.

Those two parties have benefited enormously. The U.S. has got assured access to Saudi oil and the ability to greatly influence world oil prices through effective control over the immense Saudi reserves; and the major U.S. oil companies have been guaranteed an endless stream of revenue and predominance in the world oil industry. The special U.S.–Saudi relationship has also meant that the U.S. has been able to sell billions of dollars' worth of weaponry to the Saudis, attract billions of dollars' worth of recycled Saudi money into the U.S. financial system and maintain a military presence in the kingdom—a military presence that was expanded to include more than 5,000 U.S. troops after 1991. In exchange, the family of Ibn Saud—who once reportedly carried the entire treasury of his struggling kingdom in the saddlebags of a camel—has become undoubtedly the richest family in the history of the world.

But the deal carried the seeds of future disaster; indeed, it has been central to the rise of Al Qaeda. Whatever legitimacy the royal family enjoyed in this fledgling kingdom (pulled together after tribal battles at the beginning of the twentieth century) would inevitably be undermined by the fact that it was kept in power by a foreign government—one whose culture and behaviour were largely opposed by the Saudi people. A key element in Ibn Saud's claim to legitimacy in the eyes of his people was his role as the defender of Islam, particularly the extreme fundamentalist Wahhabi sect. The Saud dynasty and the Wahhabi clerics have mutually reinforced each other's authority in the desert territory ever since a deal was struck between them in the eighteenth century. It was at the Great Mosque of Mecca in 1926 that Ibn Saud was proclaimed king and charged with protecting Islam's holy sites and with upholding Wahhabism, or *tawhid*, which became the official religious doctrine of the new kingdom. Wahhabism was openly disdainful of the self-indulgence of modern Western lifestyles— the lifestyles that America, above all, represented. But with vast oil revenues pouring into the royal treasury—and from there into the hands of some five thousand princes in the extended royal family (out of a total Saudi population of some 22 million)—the traditional Wahhabi ban on ostentatious living became more and more a theoretical concept among the upper classes of the kingdom.

It's not surprising, then, that the royal family came to be regarded by many Saudis as hypocritical and untrustworthy, as having made a lucrative pact with the devil. Indeed, not only was America synonymous with licentious modern consumerism, but perhaps more importantly, Saudis were angered by Washington's unconditional support for Israel. To a far greater extent than westerners realize, there is a burning sense of injustice and outrage throughout the Arab world over the plight of the Palestinians. "You can't imagine how deep it is in the Arab and Muslim psyche," says Syrian-born Ibrahim Hayani, who teaches economics and political science at Toronto's Ryerson University. "There isn't a single modern Arab poet who has not written about the Palestinian tragedy and the suffering of the Palestinians under the Israeli occupation." Hayani argues that Washington's unwavering support for Israel has thus put the U.S. on the side of injustice in the eyes of the Saudi population, and the royal family's failure to confront the U.S. over Israel has been a key source of the popular resentment against the U.S.-Saudi alliance.

With the royal family increasingly discredited among the Saudi populace, an elite of religious leaders became a force within the kingdom, commanding widespread allegiance and representing a challenge to the legitimacy of the Western-oriented Saudi rulers. The royal family understood the importance of keeping this religious elite onside, nervously catering to it, even sometimes at the risk of alienating Washington. (It was pressure from this religious elite, for instance, that persuaded King Faisal to impose the Saudi oil embargo against the U.S. for supporting Israel in the 1973 Arab-Israeli conflict.) The royal family lavished funds on the religious elite and its extensive system of mosques and religious schools, or *madrasahs*, which became hotbeds of Islamic fundamentalism, religious intolerance and militancy.

It's interesting to note that this fundamentalist activism only turned to terrorism after 1991, following the positioning of large numbers of U.S. troops in Saudi Arabia, something that was regarded as a sacrilege. As Elizabeth Rubin noted in an article on Wahhabi activists in *The New York Times Magazine*, "[I]t wasn't until Saddam Hussein invaded Kuwait in 1990, and President Bush vowed to drive him out and *stationed American troops on Saudi soil*, that the young men embraced political violence [italics added]." With radical sheiks denouncing the royal family for permitting infidel troops on Muslim soil, a cadre of young, religious Saudis became caught up in the call for violent action. Rubin quotes one young man, Mashari al-Thaydi, who said he decided to go for military training in Afghanistan because "[w]e all believed that the infidels were here because the Islamic world was weak . . . that talking and educating wasn't enough. We had to get physically active." In 1992, Mashari and a few others firebombed a popular video store in Riyadh, destroying shelves of American movies. Their next target was a centre that catered to widows and the poor but was seen by the militants as a vehicle for liberating women—another idea they found repugnant.

It is important to recognize the significance of the stationing of U.S. troops as a catalyst in this escalating violence. Attempts by the Saudi government to crack down proved largely unsuccessful, and by the mid-1990s Saudi Arabia was in the grips of a potent Islamist rebellion. As Rubin notes: "American soldiers still hadn't pulled out." In fact, U.S. troops remained a hugely provocative symbol until the spring of 2003, when most of them were rather abruptly withdrawn—a move that might have had more significance if U.S. troops hadn't been invading the fellow Muslim nation of Iraq at the same time.

There has been plenty of criticism in the West of the Saudi government for its failure to clamp down on this emerging terrorist movement, and also criticism of Washington for not being more insistent that the Saudis take action. But there's been less willingness to acknowledge that the Saudi–U.S. pact is itself a central part of the problem. Clearly, there is much about America that is deeply offensive to a large number of Saudis, but they have been forced to accept their nation's close political relationship with America. Saudi citizens have had no power to reject this relationship, much as they may disapprove of it, because there is no democracy in Saudi Arabia. And Washington is more than an innocent bystander in the situation. The U.S. has provided the full heft of its military might to protect the dictatorial Saudi monarchy for almost sixty years, allowing it to hold on to power in the volatile region.

Without Washington's protection, the Saudi royal family would almost certainly have been toppled by now. Indeed, it might well have been toppled decades ago by moderate reformers—like Prince Talal and Abdullah Tariki, who in the late 1950s urged democratic reforms as well as pushing for a tougher deal with Aramco. Both men were exiled; other opponents of the regime have ended up suffering harsher fates. In the face of this political reality, the only acceptable and effective vehicle for resistance became the one independent organization that has been permitted to operate: the official religion. Washington's support for the Saudi dictatorship has had the effect, then, of blocking moderate reform efforts, thereby driving opposition into the confines of religious institutions that happen to be staunchly anti-Western.

As we've seen, Washington operated in the same anti-democratic way in Iran, with equally disastrous results. Indeed, the American indifference to democracy is even more flagrant in the Iranian case, because Iran, unlike Saudi Arabia, actually was a vibrant, functioning democracy before the U.S. intervened in 1953. As in Saudi Arabia, the U.S. involvement was directly tied to oil. Washington intervened in Iran to help the major oil companies in their attempt to force the elected government of Mohammed Mossadegh to back off its decision to nationalize the Iranian oil industry. Washington then went further and actually toppled Mossadegh's government. By overthrowing Mossadegh and reinstalling the shah, Washington intervened directly *against* democracy. With Washington's support over the following two and a half decades, the shah was able to block all public dissent, which had the effect—just as in Saudi Arabia—of eliminating moderate reform efforts. Moderates came to be regarded among dissidents as naive and hopelessly ineffective, unable to stand up to the rigours of seriously challenging the ruthless shah and his American backers. As one Iranian militant later explained: "We are not liberals like . . . Mossadegh, whom the CIA can snuff out."

Thus in both Saudi Arabia and Iran, resistance coalesced in the one place where it could find a safe haven: the local mosque. Religion became a magnet not just in its own right but also as a refuge and organizational centre for resistance to the regime. The result was the emergence of a steadfast, religion-based, virulently anti-American movement bent on replacing the regimes in power, which were seen as integrally connected to Washington and all the West's corrupting influences. In Iran, the movement eventually prevailed, sweeping the shah's regime from power in 1979 in a violent revolution. In Saudi Arabia, the movement hasn't succeeded in bringing down the Saudi regime—at least not so far—but it did carry out the atrocity of 9/11. Both the Iranian revolution and the attacks of September 11 were clearly against the interests of America, and both were also clearly linked to Washington's anti-democratic actions in these countries.

Washington now loudly proclaims its desire to bring democracy to the Middle East, and there is plenty of talk about the lack of democracy there, but no talk—or even acknowledgment—that the U.S. bears a heavy share of the blame for this lack of democracy. In both Saudi Arabia and Iran, the United States has actively blocked the establishment of democracy—propping up a ruthless dictatorship in Saudi Arabia and actually overthrowing a popular, democratically elected government in Iran. (It could also be added that Washington offered crucial support for Saddam Hussein in the 1980s, allowing him to consolidate his iron control over Iraq.) The lack of democracy in the Middle East, so often lamented by commentators, is not the product of some mysterious, inexplicable Islamic sensibility, but rather, to at least some extent, the result of American actions in the region.

This unwillingness to see any connection between U.S. actions and the Middle East's democratic deficit leads Washington policymakers to conclude that its difficulty winning the hearts and minds of the region's people boils down to a public relations problem.

A bipartisan Congressional committee addressing the issue of America's growing unpopularity in the world following the Iraq war identified a few things that could be done to improve America's image, such as appointing more "public diplomacy officers," hiring more State Department officials who are fluent in Arabic and "TV ready," and encouraging private U.S. media companies to develop creative new ways to reach out to Arab youth. Similarly, Margaret D. Tutwiler, appointed by the White House to address rising hostility towards the U.S., particularly in the Muslim world, seemed to see the problem as one of sprucing up the U.S. image and improving its communications strategy. Speaking before the congressional committee in February 2004, she lamented that it was "a problem that does not lend itself to a quick fix or a single solution or a simple plan." But she thought Washington was on the right track with its $600-million budget for worldwide public diplomacy, including exchange programs, partnerships between U.S. embassies and local institutions, and textbook distributions in foreign schools. If the world is signalling its disapproval, obviously the answer lies in Washington making its case more loudly and more creatively. The one thing that is never considered is changing America's behaviour.

Two weeks after 9/11, the White House announced that President Bush was going to give a televised address to explain to a baffled, traumatized nation why America was so hated. For a moment, it almost seemed that the horrendous nature of the tragedy and the public's intense desire to make sense of it had so shaken up American politics that there was a real willingness to examine what lay behind it. Was there finally going to be some honesty about the way America had conducted itself abroad and the impact its actions might be having on the safety of Americans? Were we actually about to begin a serious search for "root causes"?

But when Bush delivered his remarks, he said nothing about the history of U.S. actions in the region, nothing about what role Washington had played in propping up hated dictators, nothing about how millions of people had been denied democracy for decades in order to accommodate the interests of U.S. oil companies. Instead, Bush said terrorists had attacked America because they hated democracy and were "jealous of our freedoms"—an explanation that, while perhaps satisfying to some, was ultimately delusional, rather like a girl convincing herself that other girls hate her because she's pretty, when really it's because they think she's a jerk.

Nothing is considered more important in the credo of the global economy than efficiency. We are told, for instance, that we must accept heavy job losses because it is more efficient for corporations to produce goods elsewhere, where labour costs are lower. We are also told that, in the name of efficiency, our tax system should give special breaks to holders of capital, to encourage the accumulation and investment of that capital. Similarly, it seems, our social programs must be lean because, we're told, making people too comfortable will make them unproductive and create inefficiencies in our economy. These sorts of policies are often advocated, despite the fact that they lead to great inequality and human suffering. No matter; efficiency is our god. How odd, then, that when it comes to the engine of the modern economy— energy—the goal of efficiency is essentially discarded.

This is especially strange since the field of energy is one where efficiency offers such promise. Whereas many of the economic policies cited in the paragraph above offer questionable efficiency gains, improvements in energy efficiency are clear-cut and significant.

Any meaningful search for solutions to our energy dilemma— dwindling worldwide supplies of oil and the threat of climate change—starts with improving our energy efficiency. By becoming more efficient in our use of energy, we are able to achieve the same results while consuming much less fuel. Yet simple solutions for making our world dramatically more energy-efficient exist and are essentially ignored by our governments. We've seen this in their refusal to impose higher fuel efficiency standards, even though the transportation sector accounts for almost half of world oil consumption. Similarly, enormous energy savings could be achieved if available technology were adopted for lighting commercial buildings. The electricity wasted through lighting systems is a major source of greenhouse gases, since a large amount of electricity is generated through the burning of coal, the most environmentally damaging fossil fuel. If all commercial buildings in the U.S. were required to install the most up-to-date lighting equipment, the rise in carbon dioxide emissions would virtually be halted.

One response to such obvious solutions is to protest that we can't afford them. Needless to say, this assertion is hopelessly short-sighted, and utterly fails to take into consideration the costs of not addressing our relentless energy consumption. But before we even get to this sort of big-picture approach, let's start by simply rejecting the assertion that being more energy efficient would be more costly. On the contrary, it would be a lot less costly, for the simple reason that energy costs money. The less of it we use, the less we have to spend. The costs of installing the most fuel-efficient technologies pay for themselves in energy cost savings, usually in very short order, and then go on to generate significant ongoing savings. As energy critic Amory Lovins has pointed out, the improvements in efficiency that could be achieved in the U.S. with existing technology amount to about $300 billion a year: "In this context, uncertainties about climate become irrelevant: we should buy energy efficiency merely to save money."

This brings us to what can only be described as the central *perversity* of energy policy: not only do we fail to do what is needed and possible and affordable—*we actively put up roadblocks to prevent what is needed, possible and affordable*. Of course, we don't admit we're doing this. Officially, the Canadian and U.S. governments—even the Cheney task force on energy—endorse energy efficiency and even put some money into its research and development. But, at the same time, they completely undermine those efforts by providing far, far more money in subsidies to the fossil fuel industry. Not only does this give the fossil fuel industry a leg up, but it means that competing alternatives are confined to the margins, apparently unfeasible and unworkable because they can't compete with fossil fuels in the marketplace (even though it's not a genuinely *competitive* marketplace). In fact, the fossil fuel industry is only "competitive" because of the enormous subsidies, most of them hidden or invisible, that it receives from our governments. In other words, we are massively subsidizing the fossil fuel industry and barely subsidizing the alternatives to it, even though the fossil fuel industry is highly destructive to the earth, and the alternatives are not. It's hard to characterize this as anything other than perverse.

This perversity can be measured because there's extensive documentation of government subsidies to different industries. Drawing on these sources, for instance, a report prepared for the Winnipeg-based International Institute for Sustainable Development showed how extensive are the subsidies to the fossil fuel industry. It found that annual subsidies to the industry amounted to about $14 billion in the U.S., $5.9 billion in Canada and a total of $59 billion in all the industrialized nations that make up the OECD. The subsidies vary from country to country and take a variety of forms, including tax credits, exemptions, deferrals, preferential rates, loans, loan guarantees, exclusions, deductions, depletion allowances and accelerated depreciation. The only other heavily subsidized industry within the energy sector, the nuclear industry, received $12 billion in annual subsidies in OECD countries. Meanwhile, alternative technologies, including geothermal and biomass energy, and wind and solar power, received relatively little government support, with the exception of hydro power. "It is fossil fuels and nuclear energy (including electricity) that receive the great bulk of energy subsidies," note the report's authors, Norman Myers, a leading environmental economist, and research associate Jennifer Kent.

The fossil fuel industry is also aided greatly by massive subsidies to the car and airline sectors that encourage these oil-fuelled forms of transportation over more sensible railroad transportation. There are a number of tax advantages that promote the use of the car (and, in the U.S., particularly promote the gas-guzzling SUV). Even more significant in sheer dollar terms are the enormous public expenditures—roughly $135 billion a year in the U.S.—on the construction and maintenance of roads. (All this more than outweighs the amount of revenue collected in gasoline taxes.) The gas-guzzling airline industry is also heavily subsidized. Airlines pay only about one-sixth of the price that motorists do for a litre of fuel, because aviation fuel is exempt from taxation. This exemption is certainly bizarre considering that aviation fuel is a particularly serious polluter. In fact, the amount of carbon dioxide emitted per airline passenger is four times higher than if the passenger had travelled the same distance by car (and twelve times higher than if the travel had been by rail).

This raises the question of whether the costs of all this pollution shouldn't really be added into the equation when we're tallying up subsidies. As Myers and Kent note, the fossil fuel industry "generates such marked pollution that some analysts consider the environmental costs of fossil fuels to be at least equal to and possibly much greater than the more conventional and recognized costs." In other words, the environmental degradation caused by the fossil fuel industry represents a cost to the public—we have to pay to repair the damage—which should really be added to the other subsidies the industry receives in calculating its total subsidy package. Myers and Kent estimate the environmental costs of the fossil fuel sector worldwide at about $200 billion a year and the environmental costs of the road transportation sector to be in the range of $380 billion. And they caution that these numbers actually significantly understate the environmental costs because, while they include the costs of oil spills, air pollution, mining tailings, acid rain, etc., they do not include the biggest environmental nightmare related to the fossil fuel sector: climate change.

The authors left global warming out of their calculations, not because it wasn't relevant or significant, but because they considered the true costs ultimately unknowable. "Regrettably, no estimate can be advanced here, not even in the form of a range, as to the size of the ultimate costs of global warming beyond preliminary assertions that it could eventually cost the United States at least 1–2 percentage points of GDP.

Extrapolated to the rest of the world, this means that the total cost could readily reach *$1 trillion per year* and probably much more . . . Suffice it to say here that global warming is far and away the greatest environmental problem we can expect within the foreseeable future [italics added]."

And there's yet another whole set of costs associated with the fossil fuel industry that haven't been factored in—such as the cost of invading and occupying Iraq. Of course, the U.S. government doesn't acknowledge that oil was a motive for the invasion. Presumably, then, the $100 billion (and rising) spent on the war and occupation should be considered a subsidy for democracy, peace and freedom in the world. Even if we assume that to be true—a rather far-fetched assumption—the cost of maintaining a huge military presence in the Gulf is clearly linked to the region's strategic importance due to its massive oil reserves, as General Anthony Zinni pointed out. By any reasonable measure, then, a significant portion of this multi-billion-dollar annual cost should be considered a form of subsidy to the fossil fuel industry. (If we were all driving fuel-cell cars, the U.S. would almost certainly maintain a much smaller military presence in the Gulf, more in keeping with its military presence in other parts of the world.) Suffice it to say that if the true costs of supporting the fossil fuel industry (direct subsidies, environmental damage, military costs) were reflected at the pump, the price of gasoline would be pro-hibitive—and railways and public transit would suddenly look like much more sensible ways to get around.

All this sheds new light on the question of the viability of many of the alternative energy options. Imagine if the tens of billions of dollars spent on the fossil fuel sector and the related road trans-portation sector were spent instead on rail transportation or on developing alternative technologies—technologies that we know are feasible but which are said to be not yet fully cost-effective in a competitive marketplace. As we've seen, the whole notion of a "competitive" marketplace is a sham. The fossil fuel industry isn't forced to compete at all. On the contrary, it is massively subsidized by governments around the world, while competing technologies are barely supported. But the subsidies for the fossil fuel industry are largely invisible, and rarely acknowledged.

As a result of our governments' policies, the public is left thinking that there is no practical choice but to rely on fossil fuels the way we do. Alternative technologies may be good and desirable, the argument goes, but they just aren't feasible (that is, "competitive") yet. Many readers must have come to this conclusion after reading *Time* magazine's "Special Report: The New Energy Crisis" in July 2003. According to the article, Bush's promise (in his January 2003 State of the Union address) to put $1.2 billion towards developing a hydrogen-powered car would probably lead to nothing. The article noted that similar promises had been made by Richard Nixon in 1974, in the wake of the first energy crisis, and by other presidents since then. The problem, the *Time* writers suggested, is that presidents announce ambitious initiatives for alternative energy technologies and even energy conservation but, after the headlines fade, largely abandon these programs, catering instead "to the demands of campaign contributors and special interests." Thus, the *Time* writers correctly identified a key aspect of the problem. But they failed to go on to point out that U.S. administrations, by catering in this way to the demands of campaign contributors and special interests, had created an incredibly lopsided set of subsidies favouring the fossil fuel industry. The article barely mentions the vast range of these subsidies. Instead, the writers allude to the difficulty of "changing anything as deep-seated as America's habits of energy use," making it sound as if the real obstacle is resistance to change on the part of the public. While it may well be difficult to get people to change old energy consumption habits, it certainly would be an awful lot easier if there weren't so many financial incentives actively enticing them to stick with the oil-powered car.

What we are dealing with is not a level playing field but a playing field tilted heavily in the direction of fossil fuels. However, what we need isn't just a level playing field (although that would be a good beginning) but a playing field *tilted heavily in the opposite direction*—towards alternative energy. Under that sort of scenario, alternative technologies could quickly be developed into practical solutions, viable in the marketplace.

Take, for instance, solar power. The cost of solar cells has already dropped dramatically, from $70 per watt of production in the 1970s to less than $4 per watt today. But it has to move down to about $1 per watt before it will be "competitive" with coal and natural gas (which are both, as we've seen, heavily subsidized). Solar power has achieved these great improvements in its price competitiveness with almost no subsidy support from government; in recent years the U.S. has spent only about $100 million a year on photovoltaic (solar) research, which is roughly the same as it spends constructing three kilometres of interstate highway. Similar cost breakthroughs have happened with wind power as well, again with little financial help from government. As of the mid-1970s, wind power cost about $1 per kilowatt hour. By 1996, that had plunged all the way down to 5¢ a kilowatt hour. The future of wind power seems impressive: Denmark is planning to meet half its electricity needs through wind power by 2030.

So how has Washington responded to these promising natural alternatives to fossil fuels? In 2003, Bush's proposed budget, while providing billions in subsidies for oil, gas, coal and nuclear energy, actually *reduced* spending on wind research by 5.5 percent, on biomass energy by 19 percent and on developing zero-energy buildings—an extremely promising solution—by 50 percent. No wonder they can't "compete."

———

We use 30 percent of all the energy . . .
That isn't bad; that is good. That means that we are the
richest, strongest people in the world and that we have
the highest standard of living in the world. That is why
we need so much energy, and may it always be that way.
—RICHARD NIXON, NOVEMBER 26, 1973

The global inequality in energy consumption, as Nixon pointed out, reflects the global inequality in living standards. Less than one-fifth of the world's people, located mostly in the industrialized West, consume the vast bulk of the world's energy and resources. Although global inequality is extreme and getting worse, it has not generally been regarded—at least in the industrialized part of the world—as a significant problem, let alone a moral dilemma of immense proportions. The fact that roughly three-quarters of the world lives in what can only be described as deprivation or semi-deprivation, and that the average Bangladeshi consumes seventy times less energy than the average American, tends to be regarded as evidence that America is doing things right, that freedom and democracy work. Attempts to do anything to address the imbalance have therefore in recent years been largely restricted to efforts to remake countries around the world in the image of America—efforts that tend to revolve around prying open their economies to foreign investment. When the poverty persists in these countries (or, often, gets worse), more openness to foreign investment is urged.

Climate change brings a whole new dimension, and a whole new level of unfairness, to this global inequality. As always, the industrialized West consumes the bulk of the world's energy, but the greenhouse gases generated by that consumption are creating a problem we all share. A Bangladeshi may consume seventy times less fuel than an American, but she won't suffer seventy times less impact from global warming. (Indeed, she may suffer more; we don't really know at this point.) Greenhouse gases aren't a local phenomenon; they become part of the overall pool of emissions that hover over the earth, potentially wreaking havoc on the earth's integrated climate system. No matter how much or how little one contributes in the way of emissions, one is equally vulnerable to the consequences of what accumulates up there.

The argument traditionally used to justify global *income* inequality doesn't stand up in the case of global *energy* inequality. Traditionally, income inequality has been rationalized on the grounds that the rich, through their entrepreneurial efforts, end up making the pie bigger for all, not just eating more of it. This has always been a highly debatable proposition, but in the case of energy consumption and its impact there is no debate: the actions of the rich are clearly hurting the poor. The high energy consumption of the rich is trapping rich and poor alike under a haze of gases altering the earth's climate. We're all in the big greenhouse together. While everyone in the world is negatively affected, it seems doubly unfair that those who didn't even get the benefits of the energy consumption end up suffering equally from its negative fallout (or likely more, since they are less equipped to protect themselves).

This isn't just another example of how the Third World gets the short end of the stick. It also goes to the heart of the question of how the problem should be dealt with, how the process of global warming is to be stopped. Logically, since the industrialized world consumes the bulk of the world's energy and emits the bulk of the world's greenhouse gases, the burden of correcting the problem should fall on the industrialized world. And this is exactly what the Kyoto accord attempts to do; its focus is on getting the 55 industrialized nations to cut back their emissions. The 132 nations of the developing world, which have contributed relatively little to the problem, have been exempted from legally binding targets in the first round of Kyoto, although they are required to develop national programs aimed at reducing emissions in order to bring them into a worldwide system, to be negotiated in future rounds.

This would seem to be a logical and fair solution to a problem largely created by us in the industrialized West. And much of the West accepted it. But the U.S., the world's biggest emitter, refused to go along. A national TV advertising campaign, funded largely by the fossil fuel industry, helped convince Americans that Kyoto unfairly picked on the United States. Designed by the same ad agency that helped scuttle Bill Clinton's plan for a national health insurance scheme, the 1997 anti-Kyoto campaign focused on the fact that the developing world was exempt from the treaty. In one advertisement, a pair of scissors cut developing countries like China, India, Algeria and Mexico out of a map of the world. A voice-over said the U.S. would be "forced to make drastic cuts in energy" under Kyoto while these other countries would not.

Left out of the ad, of course, was any hint of the fact that the problem of global warming has largely been created by the West— with the leading role played by the United States. With only 4 percent of the world's population, the U.S. is responsible for roughly 25 percent of the world's greenhouse gases. The bottom line is that we in the West have already industrialized our economies to a very high degree, we've consumed oil at a galloping pace; when it comes to energy, we in the West (Canada included) are overserved. An American consumes six times as much energy as the average world citizen. So *shouldn't* the treaty ask a lot more of the United States than it does of China, India, Algeria and Mexico, whose per capita energy consumption is a fraction of the industrial world's—and particularly America's—consumption? But in the TV ad, the U.S. comes across not as the key perpetrator of global warming, but as a victim of an unfair international campaign. There goes the Third World ganging up on America again.

The West's excessive energy consumption is aggressively defended, by those like Myron Ebell from the Competitive Enterprise Institute, as part of America's freedom. As Ebell said: "When you start limiting people's access to energy, you limit their ability to make choices. We're opposed to things that limit people's choices." Choice is, of course, almost incomprehensible to the people of the developing world—not just because they generally live under dictatorships, but because they have so little access to the world's resources. Although they make up more than two-thirds of the world's population, they consume less than one-third of its energy. Some two billion people have no electricity at all. Forced to rely instead on burning wood, charcoal or dung, usually in open stoves and firepits inside tin huts and wooden shacks, they are routinely exposed to a toxic mix of chemicals that are particularly harmful to young children—and responsible for an untold number of child deaths each year. Access to transportation of any kind is also severely limited. Millions of people, particularly women, are forced to walk long distances each day to fetch water, and then return carrying it in large jugs on their heads. This lack of access to energy traps much of the population of the developing world in lives that have changed little since feudal times. To force them to reduce their already meagre energy consumption would be to force them even further back in time.

Sadly, if there was ever a willingness to question the morality of the vastly unequal consumption of resources around the world, that moment seems to have passed. In the wake of September 11, the West's willingness even to acknowledge the problem has ebbed, putting the immorality of inequality further than ever off the agenda. The moral questions of our age are now said to revolve around terrorism and the evil it embodies. An aggressive group of neo-conservative intellectuals has seized the mantle of 9/11 to insist with new vigour on the moral rightness of America, including its right to defend its "way of life"—a way of life that depends upon a voracious appetite for energy. These neo-cons call for "moral clarity" in foreign affairs, by which they apparently mean an end to the notion that cultures and values can be different but still equally valid. Instead, they assert the *moral superiority* of America and its values, and America's right to enforce these "values" as it sees fit around the world. With this mix of hubris and self-righteousness, the neo-conservatives attempt to quash any questioning of whether America's growing command over the world's resources doesn't amount to a form of global piggery. As an angry Iraqi man, speaking to an American TV crew on the streets of Baghdad just before the U.S. invasion, put it: "Must the Americans eat every meal, must they drink every drink?"

And so it is that the world moves inexorably towards a buildup of greenhouse gases with consequences too huge to even attempt to calculate. And so it is that the world embarks on an ugly showdown between the West and the Islamic world. And so it is that, silently and far away, the majority of the world's people are left living squalid lives, trudging endless miles for water and huddling around open stoves, slowly poisoning their children with dung-burning fires in their one-room shacks.

Meanwhile, here in the "advanced world," we keep alive the dream of freedom—the dream that someday we'll be able to drive 18-wheelers that accelerate like racing cars.

Out of my way, motherfucker!

Vroooooom!

NOTES

CHAPTER ONE: Fort Knox Guarded by a Chihuahua

p. 7   "Many of the voters . . ." Juan Cole, "The Iraq Elections: First Impressions," *History News Network*," January 31, 2005. http: hnn.us/articles/9941.html

p. 9   *As a massive phalanx . . .*"War begins in a Burst of Missile Fire," *DW-World.de Deutsche Welle* (Germany's International Broadcaster), March 20, 2003.

p. 10   *But such a notion was vehemently denied . . .* James Ridgeway, "A Gusher of Propaganda," *Village Voice*, March 4, 2004.

p. 13   *After meeting with Iranian President . . .* Sarah Wagner, "Venezuela's Chavez Defends Iran's Right to Atomic Energy," March 14, 2005. www.ali2net/article 4210.html.

p. 16   *Today Canada is the largest source of oil imports . . .* The White House, *Reliable, Affordable, and Environmentally Sound Energy for America's Future: Report of the National Energy Policy Task Force* (chaired by Vice-President Dick Cheney), Washington, May 2001. Chapter 8, p. 4.

p. 16   *A three-volume report . . . The Geopolitics of Energy into the 21st Century: A Report of the CSIS Strategic Energy Initiative*, The Center for Strategic and International Studies, Washington, D.C. November 2000, Volume 3, pp. 57–58.

p. 16   *When Canada signed the North American Free Trade Agreement . . .* North American Free Trade Agreement (NAFTA) Chapter 6, Article 605. Exemption for Mexico–Annex 605.

p. 17   *The U.S. report raises this very scenario . . . The Geopolitics of Energy into the 21st Century.* p. 58.

p. 17   *With obvious satisfaction . . . The Geopolitics of Energy into the 21st Century,* p. 58.

p. 18   *China is indeed eyeing the stretch . . .* Don Gillmor, "As the Oil Runs Out: the U.S. and China square off in Alberta: what card will Ottawa play?" *The Walruns,* April 2005.

p. 18   *The need to keep Alberta's tar sands . . .* Donald G. M. Coxe, *Basic Points; Foreign Policy Priorities for Bush: Asia and Oil.* Harris Nesbitt, Harris Investment Management, Chicago, January 31, 2005.

p. 19   *As energy researcher Julian Darley. . . .* Julian Darley, *High Noon for Natural Gas* (White River Junction, Vermont: Chelsey Green Publishing Company, 2004), p. 143.

p. 20   *"Backstabbed!" was the provocative. . . .* Daily Reckoning, "Urgent Global Resources Alert," March 18, 2005. www.dailyreckoning.com

p. 21   *A report commissioned by influential Pentagon adviser . . .* Peter Schwartz & Doug Randall, "An Abrupt Climate Change Scenario and Its Implications for United States National Security," October 2003. (available through Environmental Media Service, www.ems.org. The report is discussed in Mark Townsend & Paul Harris, "Now the Pentagon tells Bush: climate change will destroy us," *The Observer,* February 22, 2004.

p. 23   *But even after a copy was leaked . . .* Mark Townsend & Paul Harris, "Now the Pentagon tells Bush: climate change will destroy us," *The Observer,* February 22, 2004.

p. 25   *Journalist George Monbiot . . .* George Monbiot, "Bottom of the Barrel," *Guardian,* December 2, 2003.

p. 26   *This notion that oil production has a "peak" . . .* Much has been written on this subject. See, for instance, Kenneth Deffeyes, *Hubbert's Peak: The Impending World Oil Shortage,* (Princeton University Press, 2001).

p. 26   *Colin Campbell, one of the world's leading geologists . . .* telephone interview, May 19, 2004. Follow-up interview, April 12, 2005.

p. 26   *Campbell's view, while rejected by many oil analysts* . . . Colin J. Campbell and Jean H. Laherrere, "The End of Cheap Oil," *Scientific American*, March 1998.

p. 27   *As the Pentagon paper on abrupt climate change* . . . Schwartz and Randall, "Abrupt Climate Change Scenario and Its Implications for United States Security."

p. 27   *Campbell's message is echoed* . . . Matthew Simmons, "Energy: a Global Outlook," Deloitte & Touche's 2004 Oil & Gas Conference, November 17, 2004. Also Matthew Simmons, "Is there an Energy Crisis?" IPAA lunch, October 11, 2004.

p. 28   *"The world's supply of oil* . . . David Frum, "Don't Worry about Running Out of Oil," *National Post*, January 13, 2005.

p. 29   *The most dramatic example has been Royal Dutch/Shell* . . . Oliver Morgan, "Shareholders want to know where Shell's reserves went," *The Observer*, January 18, 2004.

p. 29   *"The entire industry is straining* . . . Chip Cummins, "Shell forced to cut estimate of reserves for a fifth time," *Wall Street Journal*, February 6, 2005.

p. 30   *Wall Street oil analyst Fadel Gheit.* . . . Telephone interview with Fadel Gheit, April 2005.

p. 30   *The investment bank Goldman Sachs* . . . Reuters, "Goldman Sachs: Oil Could Spike to $105," March 31, 2005.

p. 30   *Jeff Rubin, chief economist at CIBC World Markets* . . . Telephone interview with Jeff Rubin, May 3, 2005. See also Jeff Rubin, "Not Just a Spike," CIBC World Markets, Occasional Report #53, April 13, 2005. www.cibcwm.com/research.

p. 31   *Although now long forgotten* . . . My account of eighteenth century developments in the textile industry is drawn largely from Gale E. Christianson, *Greenhouse: The 200-year Story of Global Warming* (Vancouver: Greystone Books, 1999), pp. 39–53.

p. 34   *A cover story on Exxon* . . ."Exxon Unleashed," *Business Week*, April 9, 2001.

p. 35   As chairman of Exxon since 1993 . . . Raymond's opposition to Kyoto has been widely reported, including in Thaddeus Herrick,

"CEO's Controversial Views Lead to Tough Summer for Exxon Mobil," *Wall Street Journal*, August 29, 2001. For a more lively account, see *A Decade of Dirty Tricks: Exxon's attempts to stop the world tackling climate change*, Greenpeace, May 2002, available at www.greenpeace.org

p. 36   *As a result, the huge growth in this SUV market* . . . Public Citizen, "New Study: SUVs Riskier to Children Than Minivan, Large and Mid-Size Cars," June 11, 2003.

p. 36   *As he told the Financial Times of London* . . . David Buchan & Sheila McNulty, "A dinosaur still hunting for growth," *Financial Times*, March 12, 2002.

p. 38   *This phrase—"an animus against humanity"* . . . author interview with Myron Ebell, Washington, D.C., March 25, 2003.

p. 38   *(Officials in the Bush administration* . . . Paul Harris, "Bush Covers Up Climate Change Research," *Observer*, September 21, 2003. Ebell's correspondence is reproduced in full in *Harper's Magazine*, May 2004. pp. 25–26.

p. 42   *Even a seasoned journalist* . . . John F. Burns, "The Road Ahead May Be Even Tougher," *New York Times*, March 7, 2004.

p. 42   *Similarly, Thomas Friedman* . . . Thomas Friedman, "Worried Optimism on Iraq," *New York Times*, September 21, 2003.

p. 43   So, for instance, General Anthony Zinni . . . quoted in Michael T. Klare, *Resource Wars* (New York: Henry Holt and Company, 2001), p. 58.

p. 44   "The oil resources of Saudi Arabia . . . memo quoted in Michael T. Klare, *Blood and Oil* (New York: Henry Holt and Company, 2004), p. 36.

p. 45   *In his task force report, Cheney acknowledged* . . . *Report of the National Energy Policy Task Force*, 2001, Chapter 1, p. 1.

p. 45   *In a speech to the London Institute* . . . cited in Kjell Aleklett, "Dick Cheney, Peak Oil and the Final Countdown," Uppsala University, Sweden, May 12, 2004. The full text of Cheney's speech was available on www. petroleum.co.uk/speeches.htm.

p. 46 *"By any estimation, Middle East oil producers . . . Report of the National Energy Policy Task Force*, Chapter 8, p. 5.

p. 46 *And, as in his speech* . . . ibid. Chapter 8, p. 5.

p. 46 *A year later, Cheney openly linked* . . . White House release, *Remarks by the Vice-President to the Veterans of Foreign Wars 103rd National Convention.* Nashville, August 26, 2002.

p. 47 *"You just write them a cheque,"* quoted in Brian Bethune, "Mad, Mad World," *Maclean's*, November 22, 2004.

CHAPTER TWO: Along Comes Iraq

p. 52 *Consider, for instance, a segment on the prewar intelligence . . . Meet the Press*, NBC, July 11, 2004.

p. 54 He issued a press release praising . . . Senator Pat Roberts, press release, March 31, 2005.

p. 54 Later that spring, the British press broke a story. . . . "The Secret Downing Street memo," *Sunday Times* (London), May 1, 2005.

p. 56 *Over in Britain, the Blair government* . . . For a detailed account of the British handling of intelligence related to Iraq, see John Cassidy, "The David Kelly Affair," *New Yorker*, December 8, 2003.

p. 58 *A comprehensive investigation* . . . Bryan Burrough, Eugenia Peretz, David Rose, David Wise, "The Path to War," *Vanity Fair*, May 2004. See also Jane Mayer, "The Manipulator: How Ahmad Chalabi Sold the War," *The New Yorker*, June 7, 2004.

p. 59 *Vincent Cannistraro, a former head* . . . quoted in Julian Borger, "White House 'exaggerating Iraqi threat,'" *Guardian*, October 9, 2002.

p. 62 *In his speech to the Veterans . . . Remarks by the Vice-President to the Veterans of Foreign War 103rd National Convention.*

p. 63 *In fact, the U.S. was Iraq's largest oil customer* . . . Robert Mabro, "Is the Widely Expected War on Iraq an Oil War?" Prof. Mabro is director of the Oxford Institute for Energy Studies and his commentary, dated February 17, 2003, appeared on the institute's website, www.oxfordenergy.org.

p. 64   *The vision of a supremely dominant America* . . . Patrick E. Tyler, "U.S. Strategy Plan Calls for Insuring No Rivals Develop A One-Superpower World," *New York Times*, March 8, 1992.

p. 65   *The PNAC's overtly militaristic and imperialistic aims* . . . see the *PNAC's Statement of Principles* . . . Project for the New American Century, *Statement of Principles*, June 3, 1997. available at PNAC website, www.newamericancentury.org.

p. 65   *In January 1998* . . . PNAC letter to Clinton, available at PNAC website.

p. 66   *In a memo drawn up in 1996* . . . Brian Whitaker, "Playing skittles with Saddam," *Guardian*, September 3, 2002. See also Robert S. Greenberger & Karby Legget, "Bush Dreams of Changing Not Just Regime but Region: A Pro-U.S., Democratic Area is Goal that has Israeli, Neoconservative Roots," *Wall Street Journal*, March 21, 2003.

p. 67   *In September 2000, just before the U.S. election* . . . *Rebuilding America's Defenses: A Report of the Project for a New American Century*, September 2000. (available on PNAC website). Pearl Harbour quote is on p. 51.

p. 68   *We now know a great deal* . . . Ron Suskind, *The Price of Loyalty* (New York: Simon & Shuster, 2004), pp. 70–86, 95–97, 183–189.

p. 70   *In Plan of Attack* . . . Bob Woodward, *Plan of Attack* (New York: Simon & Shuster, 2004), pp. 9–10.

p. 70   *Clarke's account adds* . . . Richard A. Clarke, *Against All Enemies* (New York: Free Press, 2004), pp. 30–32.

p. 73   *"When we think about scenarios* . . . author interview with Robert Ebel, Washington, D.C., November 4, 2003.

p. 73   *"All presidents refer to* . . . Lee H. Hamilton, "Can Today's Rogue States be Tomorrow's Key Energy Suppliers?," Conference on the Geopolitics of Oil in the 21st Century, Washington, D.C., December 9, 1999.

p. 74   *From his corner office* . . . author interview with Fadel Gheit, New York, January 14, 2004, also follow-up telephone interviews, May 2004 and April 2005.

p. 78   *There's something almost obscene* . . . Release, "Cheney Energy task Force Documents Feature Map of Iraqi Oilfields," Judicial Watch, July 17, 2003. Map and list of foreign suitors available at www.judicialwatch.org

p. 82   *"The U.S. Majors stand to lose* . . . "Baghdad Bazaar: Big Oil in Iraq?," Deutsche Bank (Global Oil and Gas division), October 21, 2002, p. 17.

p. 82   *James A. Paul* . . . author interview with James A. Paul, New York, January 15, 2004 and telephone interview, May 26, 2004. See also James A. Paul, "Oil in Iraq: the Heart of the Crisis," "The Iraq Oil Bonanza: Estimating Future Profits," and other articles, available at www.globalpolicy.org.

p. 83   *A GAO report in August 2003* . . . Mike Allen, "GAO Cites Corporate Shaping of Energy Plan," *Washington Post*, August 26, 2003.

p. 83   *TIME magazine reported* . . . Michael Weisskopf & Adam Zagorin, "How Bush plans the game," *TIME*, March 24, 2002.

P. 84   *As the non-partisan Washington-based* . . . Release, "A Money in Politics Backgrounder on the Energy Industry," Center for Responsive Politics, May 16, 2001, available at www.opensecrets.org

p. 84   *One intriguing piece of evidence* . . . Jane Mayer, "Contract Sport: What did the Vice-President do for Halliburton?," *The New Yorker*, February 16, 2004.

p. 85   *Mark Medish, an NSC official* . . . telephone interview with Mark Medish, April 2005. Medish also quoted in Mayer, "Contract Sport."

p. 86   *There is a similar veil of secrecy* . . . Peter Beaumont & Faisal Islam, "Carve-up of oil riches begins," *Observer*, November 3, 2002.

p. 86   *As Chalabi told the Washington Post* . . . Dan Morgan & David B. Ottoway, "In Iraqi war scenario, oil is key issue," *Washington Post*, September 15, 2002.

p. 86   *(Not surprisingly, this kind of talk* . . . Terry Macalister, "BP chief fears U.S. will carve up Iraqi oil riches," *Guardian*, October 30, 2002.

p. 87　Meanwhile, another secret meeting was . . . Thaddeus Herrick, "U.S. Oil Wants to Work in Iraq," *Wall Street Journal*, January 16, 2003.

p. 88　*The major international oil companies* . . . Michael Renner, "Post-Saddam Iraq: Linchpin of a New Oil Order," *Foreign Policy in Focus*, January 2003; available at www.fpif.org

p. 88　*"[O]ne of the goals of the oil companies* . . . Michael Tanzer, "Solidarity at the Pump: A Proposal for the Oil Exporting Nations of the Third World," *North American Congress on Latin America*, Vol. XXXIV, no. 4, January/February 2001.

p. 88　*A confidential 100-page document* . . . Neil King Jr., "Bush Officials Devise a Broad Plan for Free-Market Economy in Iraq," *Wall Street Journal*, May 1, 2003.

p. 89　*At the same time, the Pentagon was also working* . . . David Teather, "American to Oversee Iraqi Oil Industry," *Guardian*, April 26, 2003.

p. 89　"Nobody in their right mind . . . quoted in Greg Palast, "OPEC on the March," *Harper's Magazine*, April 2005. See also Greg Palast, "Secret U.S. Plans for Iraq's Oil," *BBC News World Edition*, March 17, 2005.

p. 90　*Overseen by Amy Jaffe* . . . Palast, "OPEC on the March."

p. 91　*With just days to go before Christmas 2004* . . . Emad Mekay, "U.S. to Take Bigger Bite of Iraq's Economic Pie," Inter Press Service News Agency, December 23, 2004.

CHAPTER THREE: The Man to See

p. 93　*One very practical matter was deciding which major corporation* . . . Michael Shnayerson, "The Spoils of War," *Vanity Fair*, March 7, 2005.

p. 96　*The RIO contact, had, however, attracted the attention of Judicial Watch* . . . telephone interviews and correspondence with Christopher Farrell, Judicial Watch.

p. 98   *In its May 30, 2004 issue* . . . Timothy Burger, "The Paper Trail," *Time*, May 30, 2004.

p. 98   *In late October 2004* . . . Adam Zagorin & Timothy Burger, "Beyond the Call of Duty," *Time*, October 24, 2004.

p. 100   The Iraqi rebuilding skills are most evident . . . author interview with Issam Al-Chalabi, Amman, Jordan, August 26, 2003.

p. 102   *A twenty-page analysis of Iraq's oil operations* . . . *Deutsche Bank*, p. 4.

p. 104   *Subsequent contracts were to be opened up* . . . King, "Bush officials Devise a Broad Plan for Free Market Economy in Iraq."

p. 104   *But there was no retreat* . . . Philip Thornton & Andrew Gumbel, "America Puts Iraq Up for Sale," *Independent*, September 22, 2003.

p. 104   *In its 2004 Budget Plan*, Ministry of Finance/ Ministry of Planning, Republic of Iraq [Coalition Provisional Authority], Budget Plan 2004, October 2003, available at www.cpa-iraq.org

p. 104   *Washington also took control* . . . Coalition Provisional Authority, The Development Fund for Iraq, available at www.cpa-iraq.org See also Phyllis Bennis, "The Madrid Donors Conference: A Cover for Maintaining U.S. Control," *Foreign Policy in Focus*, October 23, 2003, available at www.fpif.org/commentary

p. 105   *Altogether, these contracts are worth* . . . Neil King Jr. & Glenn R. Simpson, "Pentagon Seeks to Broaden Probe of Halliburton," *Wall Street Journal*, March 11, 2004.

p. 105   *But the contracts have accounted for nearly all of KBR's recent growth* . . . Shnayerson, "Spoils of War."

p. 105   *But there have been plenty of charges* . . . Shnayerson, "Spoils of War."

p. 108   *During his years running the Defense department* . . . Mayer, "Contract Sport."

p. 109   *During the five years Cheney served as CEO* . . . David Olive, "Dick Cheney's brilliant career," *Toronto Star*, July 26, 2002. See also Dan Briody, *The Halliburton Agenda: The Politics of Oil and Money* (Hoboken, New Jersey: John Wiley & Sons, 2004), pp. 181–237.

p. 109 *While he was still CEO of Halliburton* . . . Mayer, "Contract Sport."

p. 111 *As the late John M. Blair* . . . John M. Blair, *The Control of Oil* (New York: Vintage Books, 1978), p. 399.

p. 112 *Back in the late 1940s, Saudi Arabia's King Ibn Saud* . . . The background to the establishment of the U.S. foreign tax credit, starting with Aramco in Saudi Arabia, is documented in a number of books on the history of oil. The most detailed account, which I have relied on heavily, is in Blair's *The Control of Oil*, pp. 193–204.

p. 112 *The Venezuelan documents further showed* . . . Daniel Yergin, *The Prize: The Epic Quest for Oil, Money and Power* (New York: Touchstone Edition, 1993), p. 446.

p. 113 (*Even so, legend has it that Getty* . . . Yergin, *The Prize*, p. 444.

p. 115 *By 1973, the big five major U.S. oil companies* . . . Anthony Sampson, *The Seven Sisters: The Great Oil Companies and the World They Shaped* (New York: Viking Press, 1975), p. 111.

p. 116 *This may be partly explained by the fact* . . . Blair, *The Control of Oil*, p. 196.

p. 117 *At one of the hearings, a senator asked McGhee* . . . Yergin, *The Prize*, p. 449.

p. 117 *Anthony Sampson, author of The Seven Sisters* . . . Sampson, *The Seven Sisters*, pp. 110–111.

p. 118 *"It's always tough to sell foreign aid* . . . quoted in Tim Harper, "Burning the Midnight Oil," *Toronto Star*, October 19, 2003.

p. 120 *Back in the 1930s* . . . Blair, *The Control of Oil*, pp. 159–161.

p. 120 *Once again, we find a key official* . . . Blair, *The Control of Oil*, p. 173.

p. 121 *The absurdity of this was nicely captured* . . . Blair, *The Control of Oil*, p. 215.

p. 122 *Despite evidence in the mid-1950s* . . . Blair, *The Control of Oil*, p. 12.

CHAPTER FOUR: Revolution and Ice Cream in Caracas

p. 125 *Chavez's motorcade had barely entered Iraq* . . . Jane Arraf, "Chavez's tour of OPEC nations arrives in Baghdad," CNN.com, August 10, 2000.

p. 127 *In fact, when it came to quota-cheating* . . . Howard LaFranchi, "OPEC Hopes to Settle Price Swings," *Christian Science Monitor*, September 27, 2000. See also Richard Gott, "Oil, Chavez and the Jackal," *Guardian*, October 22, 2001; and Faisal Islam, "OPEC boss tries to pour oil on troubled waters," *Observer*, October 26, 2003.

P. 131 *Charles Shapiro, the U.S. ambassador, met with coup leaders* . . . Kim Bartley and Donnacha O'Brien, *The Revolution Will Not be Televised* (documentary film, containing footage of the 2002 coup) 2003.

p. 131 *The next day, after Latin American leaders strongly condemned* . . ."Tales from a failed coup," *The Economist*, April 29, 2002.

p. 131 *Evidence has since emerged* . . . Aram Ruben Aharonian, "Hamburgers, Cured Ham and Oil," *Proceso* (Mexican news-magazine) May 1, 2002.

p. 132 *One question is:* . . . This astute question was posed by Sarah Eisen, a 17-year-old Toronto student.

p. 132 *"The Bush administration has been invaded by madness,"* . . . author interview with Hugo Chavez, Caracas, Wednesday, March 10, 2004.

p. 135 *"None of the others got off the ground"* . . . telephone interview with Michael Tanzer, April 2004.

p. 135 *Shortly after he was first elected, Chavez* . . . For background, see Gregory Wilpert, "Collision in Venezuela," *New Left Review*, No. 21, May-June 2003; and Greg Palast, *The Best Democracy Money Can Buy* (New York: Plume Books, 2003), pp. 192–197.

p. 137 *Rafael Ramirez, minister of energy* . . . Walter Martinez, "Interview with Energy and Mines Minister Rafael Ramirez," Venezuelanalysis.com, January 27, 2004, available at www.venezuelanalysis.com

p. 138 *A former PDVSA president* . . . quoted in Karen Talbot, "Coup-Making in Venezuela: The Bush & Oil Factor," Vheadline (Venezuela's electronic news), January 27, 2004.

p. 140 *The business community quickly moved to offer its full support* . . . Mark Weisbrot, "A Split-Screen In Strike-Torn Venezuela, *Washington Post,* January 12, 2003.

P. 141 *Meanwhile the mainstream U.S. media* . . . see Antonia Zerbisias, "Venezuelan news media dissected," *Toronto Star,* September 28, 2003.

CHAPTER FIVE: From Coffins to World Destruction: The Tale of the SUV

p. 144 *Faced with a particularly bleak Christmas Eve* . . . My account of Arrhenius' contribution to the understanding of global warming is drawn largely from Christianson, *Greenhouse,* pp. 105–115.

p. 147 *Over the next hundred years* . . . *Our Common Future: The World Commission on Environment and Development* (Oxford University Press, 1987), p. 4.

p. 151 *"Without any question it's the most intense peer review* . . . author interview with Robert Watson, Washington D.C., March 25, 2003. Second author interview with Watson, Washington, November 4, 2003.

p. 153 *Nobel Prize-winning economists* . . . *The Economists' Statement on Climate Change,* Redefining Progress, Washington D.C., February 1997; available at info@redefiningprogress.org

p. 154 *The campaign orchestrated by the fossil fuel* . . . David Helvarg, "Industry's Hot Air," *Mother Jones,* November 25, 1997.

p. 155 *(Skeptic Patrick Michaels* . . . Ross Gelbspan, *The Heat is On: The Climate Crisis, the Cover-Up, the Prescription* (Reading, Massachusetts: Perseus Books, 1998), pp. 39–44, 68.

p. 155 *By May 1996* . . . Gelbspan, *The Heat is On,* pp. 76–77.

p. 156 *Elizabeth May, executive director of the Sierra Club of Canada* . . . author interview with Elizabeth May, Ottawa, December 3, 2002. Telephone interviews, May 2004.

p. 156 *A second and stronger IPCC report in 1995* . . . Gelbspan, *The Heat Is On*, p. 22.

p. 156 *"There is no debate among any statured scientist* . . . Gelbspan, *The Heat is On*, p. 22.

p. 158 *Shortly after the Bush administration took office* . . . Andrew C. Revkin, "Dispute Arises Over a Push to Change Climate Panel," *New York Times*, April 2, 2002.

p. 159 *But Watson himself was soon out the door* . . . "Climate scientist ousted," BBC, April 19, 2002.

p. 159 *Speaking in Toronto that month* . . . Peter Tabuns, "Dick Cheney's Politics: There's No Fuel Like an Old Fuel," *Globe and Mail*, May 9, 2001.

p. 159 *Alberta Premier Ralph Klein* . . . "PM wants Kyoto plan by October 8," CBC-TV news. Septeber 20, 2002.

p. 161 *Union official Fred Wilson* . . . author interview with Fred Wilson, Ottawa, December 4, 2002.

p. 161 *In June, the Globe and Mail ran a story* . . . Patrick Brethour, "Klein reverses Kyoto stand; now sees no peril to oil sands," *Globe and Mail*, June 20, 2003.

p. 162 *As Keith Bradsher has noted* . . . Keith Bradsher, *High and Mighty: SUVs: The World's Most Dangerous Vehicles and How They Got That Way* (New York: Public Affairs, 2002), pp. 5–6.

p. 163 *In an accident, an SUV is two and a half times* . . . David Friedman, *Building a Better SUV: A Blueprint for Saving Lives, Money and Gasoline.* Union of Concerned Scientists (Cambridge, Mass.: UCS Publications, 2003), p. 10.

p. 164 *More than 51,500 occupants of SUVs* . . . Friedman, *Building a Better SUV,* p. 9.

p. 165 *(Overall the transportation sector* . . . Norman Myers & Jennifer Kent, *Perverse Subsidies: How Tax Dollars Can Undercut the*

Environment and the Economy (Washington, D.C.: Island Press, 2001), p. 99. Originally published by the International Institute for Sustainable Development, Winnipeg, 1997.

p. 167 *Chrysler was a particularly notable success story* . . . Richard Byrne, "Life in the Slow Lane: Tracking Decades of Automaker Roadblocks to Fuel Economy." Unpublished paper, Union of Concerned Scientists, July 2003, p. 4

p. 167 *Chrysler president Harold R. Sperlich* . . . Sperlich and Iacocca quotes cited in Byrne, "Life in the Slow Lane," p. 4.

p. 170 *Notes David Friedman* . . . author interview with David Friedman, Washington, D.C., November 3, 2003.

p. 171 *The failure to take tougher action* . . . For the special regulatory treatment of the SUV and the 1960s trade dispute background, I have relied heavily on the detailed account in Part One of Bradsher, *High and Mighty.*

p. 174 *The fear this generated in U.S. automakers* . . . quoted in Jim Motavalli, *Forward Drive: The Race to Build "Clean" Cars for the Future* (San Francisco: Sierra Club Books, p. 41.

p. 175 *More than $9 billion was spent* . . . Bradsher, *High and Mighty,* p. 112.

p. 178 *The three API executives I met with* . . . author interview with Robert Greco, Russell O. Jones, Michael Shanahan, Washington, D.C., November 4, 2003.

p. 180 *Engineers working for this non-profit group of scientists* . . . Friedman, *Building a Better SUV,* p. 12.

p. 183 *It's interesting to note that the auto industry* . . . Bradsher, *High and Mighty,* p. 31.

p. 183 In 1966, *Henry Ford II argued* . . . quoted in Byrne, "Life in the Slow Lane," p. 3.

p. 184 *By the year 2020, the U.S. auto fleet* . . . David Friedman et al., *Drilling in Detroit: Tapping Automaker Ingenuity to Build Safe and Efficient Automobiles* (Cambridge, Mass.: UCS Publications, 2001), p. 39.

p. 185 *While other car companies* . . . Jim Motavalli, *Forward Drive: The Race to Build "Clean" Cars for the Future* (San Francisco: Sierra Club Books, 2000), pp.23–24.

p. 185 *One of the reasons SUVs have been such a source of joy to Detroit* . . . Bradsher, *High & Mighty*, pp. 85–87.

p. 186 *Although the company had more than fifty assembly-line factories* . . . Bradsher, *High & Mighty*, p. 89.

p. 186 *By 2003, George Peterson, president* . . . Jonathan Weisman, "Businesses Jump on SUV Loophole," *Washington Post*, November 7, 2003.

p. 186 *"Fortunes were waiting to be made* . . . Bradsher, *High & Mighty*, p. 92.

CHAPTER SIX: The Great Anaconda

p. 190 *People often snickered* . . . My account of Samuel Van Syckel and his dealings with Standard Oil is drawn largely from Henry Demarest Lloyd, *Wealth Against Commonwealth*; edited by Thomas Cochran (Englewood Cliffs, New Jersey: Prentice-Hall, 1963), pps. 73–78.

p. 195 *Yergin portrays Rockefeller in a light* . . . Yergin, *The Prize*, p. 55.

p. 197 *There's no question that Rockefeller was a self-made man* . . . My account of Rockefeller and his involvement with the South Improvement Company is drawn from Ida M. Tarbell, *The History of the Standard Oil Company*; edited by David M. Chalmers (New York: Norton, 1969), as well as from Lloyd, *Wealth Against Commonwealth* and Yergin, *The Prize*.

p. 199 *The contracts establishing the arrangement* . . . quoted in Tarbell, *History*, p. 31.

p. 200 *"You see this scheme is bound* . . . quoted in Tarbell, *History*, p. 32.

p. 202 *It had the highest per capita income in Canada* . . . "Petrolia" in *The Canadian Encyclopedia*, available at www.thecanadianencyclopedia. com

p. 202 *The town of Pithole was said in the early days* . . . Sampson, *The Seven Sisters*, p. 19.

p. 203   *The Oil City Derrick was* . . . Tarbell, *History*, p. 47.

p. 209   *Woodrow Wilson stated the problem bluntly* . . . quoted in Richard Hofstadter, *The Age of Reform; from Bryan to F.D.R.* (London: J. Cape, 1968), p. 231.

p. 210   *Where was all this leading?* . . . Hofstadter, *Age of Reform*, p. 136.

p. 210   *"It has revolutionized the way of doing business* . . . quoted in Sampson, *The Seven Sisters*, p. 25.

p. 210   *"There is a sense in which in our day the individual* . . . Woodrow Wilson, *The New Freedom: a call for the emancipation of the generous energies of the people* (Englewood Cliffs, New Jersey: Prentice-Hall, 1961), p. 20.

p. 210   "There was a time when corporations . . . quoted in Hofstadter, *Age of Reform*, p. 224.

p. 211   *Wilson characterized the struggle* . . . Wilson, *The New Freedom*, p. 6.

p. 211   *In a line that is striking to read today* . . . Wilson, *The New Freedom*, p. 15.

p. 212   *"Nobody was under any illusion* . . . My account of the U.S. case against Rockefeller is drawn from David Bryn-Jones, *Frank B. Kellogg, a biography* (New York: Putnam, 1937), Yergin, *The Prize* and Sampson, *Seven Sisters*.

p. 212   *"With public attention riveted on the courtroom* . . . quoted in Yergin, *The Prize*, p. 108.

p. 213   *"No disinterested mind* . . . quoted in Sampson, *The Seven Sisters*, p. 28.

p. 213   *"Public opinion and the American political system* . . . Yergin, *The Prize*, p. 110.

p. 214   *The ongoing role of the Rockefeller family* . . . The holdings of the Rockefeller family are set out in the 1940 study by the U.S. Securities and Exchange Commission, *The Distribution of Ownership in the 200 Largest Nonfinancial Corporations*, p. 127. They are cited in Blair, *The Control of Oil*, p. 149.

p. 215   *Many decades after the Supreme Court's ruling* . . . quoted in Blair, *The Control of Oil*, p. 398.

p. 216 *As economist Michael Tanzer notes* . . . Michael Tanzer, "Solidarity at the Pump."

CHAPTER SEVEN: How Did Our Oil Get Under Their Sand?

p. 218 *"You didn't have to be very smart to see it"* . . . author interview with Ali A. Attiga, Amman, Jordan, August 27, 2003.

p. 219 *Hammer had recently decided to enter the oil business* . . . quoted in Robert Engler, *The Brotherhood of Oil: Energy Policy and the Public Interest* (Chicago: University of Chicago Press, 1977), p. 18.

p. 220 *As one top U.S. oil adviser later put it* . . . Walter Levy, oil adviser to the U.S. State Department, quoted in Pierre Terzian, *OPEC: The Inside Story*; translated by Michael Pallis (London: Zed Books, 1985), p. 113.

p. 222 *Edmonton economist and consultant* . . . author interview with Mark Anielski, Edmonton, September 7, 2003. See also Anielski, "Could we ever match Norway?" *Edmonton Journal*, March 26, 2002, and *Giving Away the Alberta Advantage*, Parkland Institute, Edmonton, November 1999.

p. 223 *Effectively reduced to the status of a British protectorate* . . . My account of Mohammed Mossadegh and the rise of democracy in Iran in the early 1950s is drawn from a number of sources, but particularly from Stephen Kinzer, *All the Shah's Men: An American Coup and the Roots of Middle East Terror* (Hoboken, New Jersey: John Wiley & Sons, 2003). See also Fred Halliday, *Arabia Without Sultans* (Middlesex: Penguin Books, 1975) pp. 466- 490.

p. 225 *"Our authority throughout the Middle East* . . . quoted in Yergin, *The Prize*, p. 465.

p. 225 *U.S. Secretary of State Dean Acheson* . . . quoted in Blair, *The Control of Oil*, p. 78.

p. 225 *In his account of the Iranian nationalization* . . . Yergin, *The Prize*, p. 463.

p. 226 *In May 1951, the U.S. State Department issued* . . . cited in Blair, *The Control of Oil*, p. 79.

p. 226 *"The embargo . . . was very effective due to the co-operation . . .* quoted in Blair, *The Control of Oil,* p. 79.

p. 227 *On a trip to Cairo in the fall of 1951 . . .* Kinzer, *All the Shah's Men,* p. 131.

p. 227 *British Defence Minister Emmanuel Shinwell . . .* quoted in Yergin, *The Prize,* p. 458.

p. 227 *"Holding the line against Third World nationalism . . .* Kinzer, *All the Shah's Men,* p. 131.

p. 227 *Kinzer reports that on June 14, 1953 . . .* Kinzer, *All the Shah's Men,* pp. 161–162.

p. 228   *Kinzer describes their clandestine meetings . . .* Kinzer, *All the Shah's Men,* pp. 4–16 and 191–192.

p. 228 *(As for Kermit Roosevelt . . .* Kinzer, *All the Shah's Men,* p. 202.

p. 230 *One of the most striking things about the international oil order . . .* My account of the operation of the international oil order is drawn from a number of sources, but particularly from Blair, *The Control of Oil,* Sampson, *The Seven Sisters* and Yergin, *The Prize.*

p. 232 *An internal Exxon memo . . .* cited in Blair, *The Control of Oil,* p. 36.

p. 232 *(The Saudi king's dislike of all things British . . .* Yergin, *The Prize,* p. 415.

p. 233 *A faction on SoCal's board argued forcefully . .* Blair, *The Control of Oil,* pp. 38–42.

p. 234 *Yergin says it was because the merger was favoured . . .* Yergin, *The Prize,* p. 413.

p. 236 *The agreements are astonishing in their detail . . .* for details on the agreements, see Blair, *The Control of Oil,* pp. 56–70.

p. 237 *The case involved deliberations among the majors . . .* Blair, *The Control of Oil,* p. 63.

p. 238 *"Ever since the Achnacarry agreement . . .* Tanzer, "Solidarity at the Pump."

p. 239   *Exxon vice-president Howard Page* . . . quoted in Sampson, *The Seven Sisters*, p. 168.

p. 240   *According to an intelligence report cited at U.S. Senate hearings* . . . Blair, *The Control of Oil*, p. 85.

p. 240   *IPC was up to the same sort of tricks* . . . Blair, *The Control of Oil*, p. 82.

p. 240   *Andreas Lowenfield, a legal adviser* . . . Blair, *The Control of Oil*, p.86.

p. 241   *But Washington was not deterred* . . . Blair, *The Control of Oil*, p. 87.

p. 242   *Exxon's Howard Page admitted to a U.S. Senate commission* . . . Terzian, OPEC, p. 104.

p. 242   *Issam al-Chalabi, the former oil minister* . . . author interview.

p. 243   *Yet as John Blair noted* . . . Blair, *The Control of Oil*, p. 101. The overall supply referred to by Blair is the total produced by eleven leading oil-producing nations, representing 85 per cent of the world supply, but excluding the U.S., Canada and Russia.

p. 243   *"Historically, the largest profits* . . . Tanzer, "Solidarity at the Pump."

p. 243   *John Warder, the chairman of the majors' consortium* . . . Blair, *The Control of Oil*, p. 48.

p. 244   *One can even admire what oil analyst Paul Frankel* . . . quoted in ibid, p. 52.

p. 246   *This was acknowledged by James Akins* . . . Terzian, OPEC, p. 119.

p. 247   *He also made a trip to New York to ask Exxon chairman Kenneth Jamieson* . . . Yergin, *The Prize*, p. 579.

p. 247   *As Qaddafi noted* . . . quoted in Yergin, *The Prize*, p. 578.

p. 247   *Libya's revenues shot up overnight* . . . Terzian, OPEC, p. 122

CHAPTER EIGHT: The Harem Takes on the Sisters: the Rise of OPEC

p. 249   *OPEC was the brainchild of two men* . . . For background on the two founders of OPEC, see Terzian, *OPEC*, pp. 65–96. See also Yergin, *The Prize.*

p. 251   *As a result, the companies paid very low royalty rates* . . . Terzian, *OPEC*, p. 69.

p. 251   *In fact, the bill had been drawn up* . . . Terzian, *OPEC*, p. 69.

p. 251   *"The manner in which they have exploited the wealth* . . . Terzian, *OPEC*, p. 69.

p. 254   *As we saw earlier, the majors had pushed for such regulation* . . . For historical background on the Texas Railroad Commission's role in regulating oil production, see Blair, *The Control of Oil*, pp. 159–166.

p. 255   *Howard Page, Exxon's long-time chief Middle East strategist* . . . quoted in Terzian, *OPEC*, p. 85.

p. 257   *Relations between Tariki and Aramco* . . . Terzian, *OPEC*, p. 89.

p. 261   *A jubilant Perez Alfonzo declared* . . . quoted in Sampson, *The Seven Sisters*, p. 156.

p. 261   *Tariki's firing marked the end of a brief flirtation* . . . For a discussion of the reform effort led by Prince Talal and Tariki, see Halliday, *Arabia without Sultans*, pp. 65–67, Terzian, *OPEC*, p. 88, and John Rossant, "The Return of Saudi Arabia's Red Prince," *Asia Times*, March 19, 2002.

p. 262   *James Akins, former U.S. ambassador to Saudi Arabia* . . . author interview with James Akins.

p. 262   *At a packed gathering of the Arab Oil Congress* . . . Terzian, *OPEC*, p. 91.

p. 263   *"At the end, he was excluded from all government decision-making* . . . author interview with Hugo Chavez.

p. 264   *Passivity was not something that ever sat well* . . . quoted in Terzian, *OPEC*, p. 83.

p. 267   *The Arab world had been producing 20 million barrels* . . . Yergin, *The Prize*, p. 614.

p. 268   *In 1973–the year OPEC unilaterally imposed* . . . Sampson, *The Seven Sisters*, p. 266.

p. 268   *One can best appreciate the magnitude of Exxon's gain* . . . Blair, *The Control of Oil*, p. 308.

p. 268   *Daniel Yergin seems to attribute* . . . Yergin, *The Prize*, p. 659.

p. 269   *John Blair, the anti-trust economist* . . . Blair, *The Control of Oil*, p. 289.

p. 269   *The Petroleum Economist noted early in 1975* . . . cited in Blair, *The Control of Oil*, p. 290.

p. 269   *Blair concludes that, with OPEC's lack of competence* . . . Blair, *The Control of Oil*, p. 291.

p. 270   *The role played by the companies* . . . Blair, *The Control of Oil*, p. 292.

p. 271   *U.S. President Gerald Ford* . . . quoted in Sampson, *The Seven Sisters*, p. 283.

p. 272   *Washington initially sent two high-level officials* . . . Stephen Pelletiere, *Iraq and the International Oil System*, (Westport, Conn.: Praeger, 2001), p.106.

p. 273   *"Rarely, perhaps never before* . . . quoted in Terzian, *OPEC*, p. 203.

p. 273   *Average incomes in those four Gulf states* . . . Blair, *The Control of Oil*, p. 274.

p. 273   *The Economist informed its readers* . . . Terzian, *OPEC*, p. 204.

p. 274   *One analysis by the Bank of Chicago* . . . Terzian, *OPEC*, p. 207.

p. 275   *"This produced huge trade deficits for the developing countries* . . . Ali A. Attiga, *The Arabs and the Oil Crisis 1973–1986* (Safat, Kuwait: Organization of Arab Petroleum Exporting Countries, 1987), p. 5.

p. 276   *More recently, throughout the 1990s, the UN Security Council* . . . Ramsay Clark's Letter to the UN Security Council, August 27, 1999. available at www.transantional.org.

p. 277   *In 1974, the developing countries faced* . . . Blair, *The Control of Oil*, p. 274.

p. 278 *Henry Kissinger, of all people* . . . quoted in Terzian, *OPEC*, p. 210.

p. 278 *Despite promises to set aside* . . . Terzian, *OPEC*, p. 210.

p. 279 *"In the past we spoke and nobody listened* . . . quoted in Terzian, *OPEC*, p. 212.

p. 282 *Faisal was also a committed nationalist* . . . Terzian, *OPEC*, p. 235.

p. 282 *Unlike just about everyone else in the top levels* . . . Robert Baer, *Sleeping with the Devil: How Washington Sold Our Soul for Saudi Crude* (New York: Crown Publishers, 2003), p. 171.

p. 283 *When the Saudis again resisted a major price hike* . . . Terzian, *OPEC*, p. 254.

p. 284 *This prompted an angry response from even the Saudis* . . . Terzian, *OPEC*, p. 262.

p. 285 *As Yergin points out* . . . Yergin, *The Prize*, p. 718.

p. 285 *Ralph Torrie, an Ottawa energy consultant* . . . Ralph Torrie, Richard Parfett & Paul Steenhof, *Kyoto and Beyond*, David Suzuki Foundation, October 2002, pp.22- 23.

p. 286 *By August 1974, only a few months after the oil embargo* . . . Sampson, *The Seven Sisters*, p. 268.

CHAPTER NINE: King of the Vandals

p. 287 *In a sun-filled restaurant* . . . author interview with James Akins, Washington, D.C., March 25, 2003. Second interview, Washington, D.C., November 3, 2003.

p. 291 *Daniel Yergin notes that* . . . Yergin, *The Prize*, p. 395.

p. 291 *As Herbert Feis, economic adviser* . . . quoted in Yergin, *The Prize*, p. 396.

p. 292 *"The military demands for an operation of this size* . . . quoted in David Isenberg, "The Rapid Deployment Force: The Few, the Futile, the Expendable," Cato Policy Analysis No. 44, Cato Institute, Washington, D.C., November 8, 1984.

p. 293 *Asked if Washington would consider* . . . "Kissinger on Oil, Food, and Trade," *Business Week*, January 13, 1975.

p. 293 *Rather, President Gerald Ford* . . . "Gerald Ford: They Will See Something Is Being Done," *Time*, January 20, 1975.

p. 293 *Secretary of Defense James R. Schlesinger* . . . "Now a Tougher U.S.: Interview with James R. Schlesinger, Secretary of Defense," *U.S. News & World Report*, May 26, 1975.

p. 293 *James H. Noyes, deputy assistant defense secretary* . . . James H. Noyes, *The Clouded Lens: Persian Gulf Security and U.S. Policy* (Stanford: Hoover Institution Press, 1979), p. 91.

p. 294 *"What we decided* . . . quoted in Klare, *Resource Wars*, p. 60.

p. 295 *"Americans," observed New York Times correspondent* . . . Kinzer, *All the Shah's Men*, p. 202.

p. 295 *"[I]t was evident to senior policy makers* . . . Klare, *Resource Wars*, p. 60.

p. 295 *In January, 1980, President Jimmy Carter* . . . Jimmy Carter, *State of the Union Address*, January 21, 1980.

p. 296 *By 1981, Washington's long-standing preoccupation* . . . quoted in Isenberg, "The Rapid Deployment Force."

p. 296 *(The strait has been described* . . . Klare, *Resource Wars*, p. 73.

p. 296 *In one 1981 military exercise* . . . Isenberg, "The Rapid Deployment Force."

p. 297 *Qasim went on to be a fairly popular leader* . . . Pelletiere, *Iraq and the International Oil System*, pp. 123–133.

p. 298 *There are credible reports of CIA involvement* . . . Pelletiere, *Iraq and the International Oil System*, pp. 131–135.

p. 298 *"In the eyes of the companies* . . . Pelletiere, *Iraq and the International Oil System*, p. 143.

p. 299 *The U.S. support for Saddam* . . . Christopher Marquis, "U.S. courted Iraq despite arms use," *New York Times*, December 23, 2003.

p. 300 *Rumsfeld's second trip to Baghdad* . . . Dana Priest, "Rumsfeld visited Baghdad in 1984 to Reassure Iraqis, Documents Show," *Washington Post*, December 19, 2003.

p. 301   *According to Pelletiere, Saddam "wanted to preclude* . . . Pelletiere, *Iraq and the International Oil System*, p. 165.

p. 301   *By 1989, General Colin Powell* . . . Klare, *Resource Wars*, p. 61.

p. 302   *Kuwait showed its gratitude* . . . Pelletiere, *Iraq and the International Oil System*, p. 216.

p. 302   *According to an official Iraqi transcript of the encounter* . . . Excerpts from an Iraqi Document on Meeting with U.S. envoy," *New York Times*, September 23, 1990.

p. 303   *British journalists got a copy of it* . . . Martin O'Malley & Owen Wood, "Bombs over Baghdad: 10 years after Desert Storm," Canadian Broadcasting Corporation Online, January 1, 2001. See also Carleton Cole, "US Ambassador to Iraq April Glaspie," *Christian Science Monitor*, May 27, 1999.

p. 303   *In 1992, presidential challenger Ross Perot* . . . Russell Dybvik, "Allegations on Communications with Iraq Denied," United States Information Agency, October 20, 1992.

p. 304   *"The U.S. will continue to use a variety* . . . quoted in Klare, *Resource Wars*, p. 62. See also Robert Dreyfuss, "The Thirty Year Itch," *Mother Jones*, March/April 2003.

p. 305   *Washington also provided more than $42 billion* . . . Klare, *Resource Wars*, p. 64.

p. 305   *This focus on oil was acknowledged* . . . The Geopolitics of Energy into the 21st Century, Vol. 3, p. 3.

p. 306   *Michael Ignatieff, a Canadian academic* . . . Michael Ignatieff, "Second, sober thoughts," *Toronto Star*, December 26, 2003.

p. 307   *By 2010, an estimated 95 percent* . . . Michael Meacher (former British environment minister), "This war on terrorism is bogus," *The Guardian*, September 6, 2003.

p. 308   *"The U.S. views oil as a key weapon* . . . author interview with Michael Tanzer.

p. 309   *As violence escalated in the fall of 2003* . . . Thomas Friedman, "It's No Vietnam," *New York Times*, October 30, 2003.

p. 309 *Bob Woodward has helped flesh out this portrait* . . . quoted in Paul Koring, "Book says Bush consumed with destroying Hussein," *The Globe and Mail*, April 21, 2004.

p. 310 *Addressing the National Endowment for Democracy* . . . White House release, "President Bush Discusses Freedom in Iraq and Middle East," Washington, November 6, 2003.

p. 310 *The following month, Donald Rumsfeld visited Baku* . . . Editorial, "Our Man in Baku," *Washington Post*, January 25, 2004.

CHAPTER TEN: Vroooooom!

p. 315 *Bin Laden refers to "eighty years of humiliation* . . . *Los Angeles Times*, October 7, 2001.

p. 316 *"If Iraq did not have oil* . . . author interview with Fadel Gheit.

p. 316 *That was the beginning of the "special relationship"* . . . See Halliday, *Arabia Without Sultans*, pp. 47–59, George Lenczowski, *The Middle East in World Affairs*, fourth edition (Ithaca: Cornell University Press, 1980), pp. 583–586, Yergin, *The Prize*, pp. 402–405, and Daniel Fisher, "The Prize: The war against Osama bin Laden is also a contest for the world's largest producer of oil," *Forbes Magazine*, November 12, 2001.

p. 317 *In exchange, the family of Ibn Saud* . . . Yergin, *The Prize*, p. 284.

p. 319 *As Elizabeth Rubin noted in an article* . . . Elizabeth Rubin, The Opening of the Wahhabist Mind, *New York Times Magazine*, March 7, 2004.

p. 321 *As one Iranian militant later explained* . . . Ayatollah Ali Khamenei, quoted in Kinzer, *All the Shah's Men*, p. 203.

p. 322 *A bipartisan Congressional committee* . . . Christopher Marquis, "U.S. Image Abroad Will Take Years to Repair, Official Testifies," *New York Times*, February 5, 2004.

p. 323 *Instead, Bush said terrorists had attacked America* . . . This apt analogy comes from Toronto writer Barbara Nichol.

p. 325 *As energy critic Amory Lovins has pointed out* . . . cited in Myers & Kent, *Perverse Subsidies*, op. cit., p. 34. See also Paul Hawken,

Amory Lovins & Hunter Lovins, *Natural Capitalism: Creating the Next Industrial Revolution* (Boston: Little, Brown & Company, 1999), and Andrew Heintzman & Evan Solomon (eds.) *Fueling the Future: How the Battle Over Energy is Changing Everything* (Toronto: House of Anansi Press, 2003).

p. 326 *Drawing on these sources, for instance* . . . Myers & Kent, *Perverse Subsidies*, p. 85.

p. 326 *The subsidies vary from country to country* . . . Myers & Kent, *Perverse Subsidies*, p. 65.

p. 327 *The authors left global warming out of their calculations* . . . Myers & Kent, *Perverse Subsidies*, pp. 82–83.

p. 329 *Many readers must have come to this conclusion* . . . Donald L. Bartlett & James B. Steele, "Special Report: The New Energy Crisis," *Time*, July 21, 2003.

p. 330 *Take, for instance, the cost of solar power* . . . Myers & Kent, *Perverse Subsidies*, pp. 89–90.

p. 330 *In 2003, Bush's proposed budget* . . . Bartlett & Steele, "Special Report: The New Energy Crisis."

p. 331 *The fact that roughly three-quarters of the world* . . . Myers & Kent, *Perverse Subsidies*, p. 63.

ACKNOWLEDGEMENTS

The idea for this book came from my long-time and much-valued friend, Gord Evans, who suggested it over a pleasant outdoor lunch. It took only a few minutes for me to realize it was a great idea, without fully realizing at that point just how central the story line would become to perhaps the major political narrative of our times.

I am indebted to a number of people who generously gave their time and effort to assist me with this project, and whose expertise greatly helped me in understanding the subject: notably Ralph Torrie, Elizabeth May, Mark Anielski, Robert Engler, Ibrahim Hayani, Maria Victor and Ricardo Salame.

I want to thank the very capable, creative crew at Doubleday, starting with publisher Maya Mavjee, who was constant and enthusiastic in her support and always a real pleasure to deal with. It's also been great working with trusted old friends like Brad Martin and Scott Sellers, with new ones like Nick Massey-Garrison, Scott Richardson, Lara Hinchberger and Christine Innes, as well as with freelance copy editor John Sweet. I'm also appreciative of the talents of my agent, Bruce Westwood.

As with all my previous books, I was privileged to have as my editor David Kilgour, whose keen sense of structure—and humour—made editing sessions painless. Thanks also to Barbara Nichol for her thoughtful input into the editing process. (While

I'm at it, I'd also like to thank her for extremely helpful suggestions on my columns for the *Toronto Star*, which I send to her each week before submitting. Invariably, she figures out how to improve them, often quite a bit.)

And thanks to the cast of great characters in my personal life: my parents, Audrey and Jack McQuaig, my brothers Peter, Don and John McQuaig, my sister Wendy Fallis, my in-laws Diane McQuaig, Janet Allen and Fred Fallis, as well as: Fred Fedorsen, Tasha Fedorsen, Tom Walkom, David Cole, Linda Diebel, Kenny Finkleman, Ellen Vanstone, Peter Duffin, Baghya Patel, Cora Ruffule and Stephanie Chan. I also want to thank Andrzej Tarnas for all the wonderful hours we've spent together in the garden.

As always, my final thanks goes to my terrific, smart, loving daughter Amy.

INDEX

ABOUT THE AUTHOR

Journalist and best-selling author **LINDA McQUAIG** has developed a reputation for challenging the establishment. Winner of a National Newspaper Award and an Atkinson Fellowship for Journalism in Public Policy, she has written for *The Globe and Mail*, the *National Post*, the *Montreal Gazette*, and numerous national magazines. She now writes a weekly political column on the op-ed page of the *Toronto Star*. She is author of six previous books—all national best-sellers—on politics and economics, including *Shooting the Hippo*, *The Cult of Impotence* and, most recently, *All You Can Eat: Greed, Lust and the New Capitalism*.